I0064921

Introduction to Remote Sensing

Introduction to Remote Sensing

Edited by Max Hopkins

SYRAWOOD
PUBLISHING HOUSE

New York

Published by Syrawood Publishing House,
750 Third Avenue, 9th Floor,
New York, NY 10017, USA
www.syrawoodpublishinghouse.com

Introduction to Remote Sensing
Edited by Max Hopkins

© 2018 Syrawood Publishing House

International Standard Book Number: 978-1-68286-604-7 (Hardback)

This book contains information obtained from authentic and highly regarded sources. All chapters are published with permission under the Creative Commons Attribution Share Alike License or equivalent. A wide variety of references are listed. Permissions and sources are indicated; for detailed attributions, please refer to the permissions page. Reasonable efforts have been made to publish reliable data and information, but the authors, editors and publisher cannot assume any responsibility for the vailidity of all materials or the consequences of their use.

Trademark Notice: Registered trademark of products or corporate names are used only for explanation and identification without intent to infringe.

Cataloging-in-Publication Data

Introduction to remote sensing / edited by Max Hopkins.
   p. cm.
Includes bibliographical references and index.
ISBN 978-1-68286-604-7
1. Remote sensing. 2. Space optics. I. Hopkins, Max.
G70.4 .I58 2018
621.367 8--dc23

TABLE OF CONTENTS

PREFACE

Remote sensing refers to the process of acquiring information of objects without physically examining the object. It is an important process used in fields like ecology, geography, geology, land surveying, glaciology, hydrology and oceanography, etc. It generally includes satellites and aircraft based sensors to root out information about objects present in the deep sea, atmosphere, etc. This book presents the complex subject of remote sensing in the most comprehensible and easy to understand language. It includes a detailed explanation of the various concepts and applications of the field. Some of the diverse topics covered in it address the varied branches that fall under this category. This textbook will serve as a valuable source of reference for those interested in remote sensing.

Given below is the chapter wise description of the book:

Chapter 1- The acquiring of information about distant objects and surfaces without physically coming in contact with them is known as remote sensing. It can be categorized into active and passive remote sensing. Some of the examples of remote sensing include photography, RADAR and LiDAR. The data derived from remote sensing can be used for developing a better understanding of areas which are remote. This chapter will provide an integrated understanding of remote sensing.

Chapter 2- Remote sensing mainly uses technology which helps in achieving a global coverage. The different types of satellites used in orbits are geosynchronous orbit, polar orbit and sun-synchronous orbit. These remote sensing satellites are used to assure that we receive the world's coverage everyday. This section provides a plethora of interdisciplinary topics for better comprehension of the technologies of remote sensing.

Chapter 3- Digital elevation model (DEM) is the model that is used to represent the surface of a planet in a digital plane. They can be of two types- raster or vector based triangular irregular network. Digital elevation modelling is mostly done with the help of remote sensing and rarely done by using data collected directly. This chapter discusses the methods of digital elevation modelling in a critical manner providing key analysis to the subject matter.

Chapter 4- Remote sensing is applied in various fields. Some of these areas are flood mapping, environmental monitoring, irrigation management and watershed. Remote sensing helps in generating awareness related to social and economic problems like deforestation, land usage and discovering natural resources. The topics discussed in the section are of great importance to broaden the existing knowledge on remote sensing.

Chapter 5- Microwave remote sensing is the ideal method that can be used to study atmospheric activity since microwaves can penetrate clouds, rain, light and other atmospheric phenomena. The chapter strategically encompasses and incorporates the advances in the field of remote sensing.

At the end, I would like to thank all those who dedicated their time and efforts for the successful completion of this book. I also wish to convey my gratitude towards my friends and family who supported me at every step.

Editor

Basics of Remote Sensing

The acquiring of information about distant objects and surfaces without physically coming in contact with them is known as remote sensing. It can be categorized into active and passive remote sensing. Some of the examples of remote sensing include photography, RADAR and LiDAR. The data derived from remote sensing can be used for developing a better understanding of areas which are remote. This chapter will provide an integrated understanding of remote sensing.

Remote Sensing

Remote sensing is an art and science of obtaining information about an object or feature without physically coming in contact with that object or feature. Humans apply remote sensing in their day-to-day business, through vision, hearing and sense of smell. The data collected can be of many forms: variations in acoustic wave distributions (e.g., sonar), variations in force distributions (e.g., gravity meter), variations in electromagnetic energy distributions (e.g., eye) etc. These remotely collected data through various sensors may be analyzed to obtain information about the objects or features under investigation. In this course we will deal with remote sensing through electromagnetic energy sensors only.

Thus, remote sensing is the process of inferring surface parameters from measurements of the electromagnetic radiation (EMR) from the Earth's surface. This EMR can either be reflected or emitted from the Earth's surface. In other words, remote sensing is detecting and measuring electromagnetic (EM) energy emanating or reflected from distant objects made of various materials, so that we can identify and categorize these objects by class or type, substance and spatial distribution.

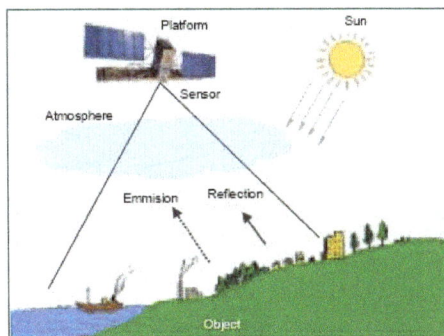

Schematic representation of remote sensing technique

Remote sensing provides a means of observing large areas at finer spatial and temporal frequencies. It finds extensive applications in civil engineering including watershed studies, hydrological states and fluxes simulation, hydrological modeling, disaster management services such as flood and drought warning and monitoring, damage assessment in case of natural calamities, environmental monitoring, urban planning etc.

Basic concepts of remote sensing are introduced below.

Electromagnetic Energy

Electromagnetic energy or electromagnetic radiation (EMR) is the energy propagated in the form of an advancing interaction between electric and magnetic fields (Sabbins, 1978). It travels with the velocity of light. Visible light, ultraviolet rays, infrared rays, heat, radio waves, X-rays all are different forms of electro-magnetic energy.

Electro-magnetic energy (E) can be expressed either in terms of frequency (f) or wave length (λ) of radiation as

$$E = h\,c\,f \qquad \text{or} \qquad h\,c\,/\,\lambda$$

where h is Planck's constant (6.626 x 10^{-34} Joules-sec), c is a constant that expresses the celerity or speed of light (3 x 10^8 m/sec), f is frequency expressed in Hertz and λ is the wavelength expressed in micro meters (1μm = 10^{-6} m).

As can be observed from equation (1), shorter wavelengths have higher energy content and longer wavelengths have lower energy content.

Distribution of the continuum of energy can be plotted as a function of wavelength (or frequency) and is known as the EMR spectrum.

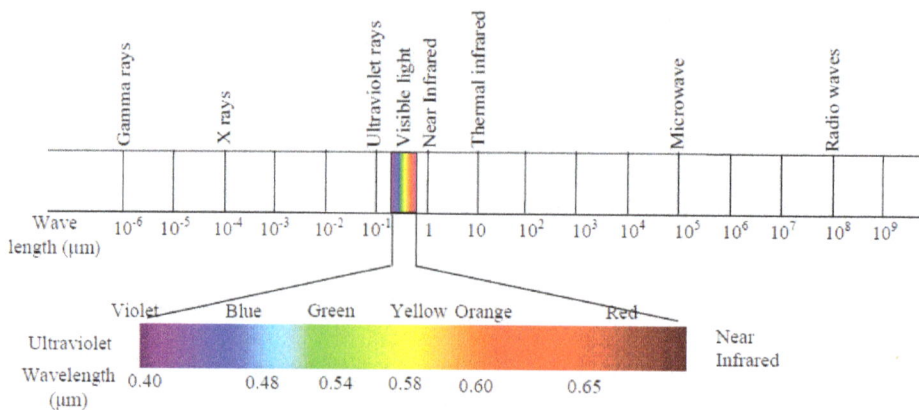

Electromagnetic radiation spectrum

In remote sensing terminology, electromagnetic energy is generally expressed in terms of wavelength, λ.

All matters reflect, emit or radiate a range of electromagnetic energy, depending upon the material characteristics. In remote sensing, it is the measurement of electromagnetic radiation reflected or emitted from an object, is the used to identify the target and to infer its properties.

Principles of Remote Sensing

Different objects reflect or emit different amounts of energy in different bands of the electromagnetic spectrum. The amount of energy reflected or emitted depends on the properties of both the material and the incident energy (angle of incidence, intensity and wavelength). Detection and discrimination of objects or surface features is done through the uniqueness of the reflected or emitted electromagnetic radiation from the object.

A device to detect this reflected or emitted electro-magnetic radiation from an object is called a "sensor" (e.g., cameras and scanners). A vehicle used to carry the sensor is called a "platform" (e.g., aircrafts and satellites).

Main stages in remote sensing are the following.

A. Emission of electromagnetic radiation

- The Sun or an EMR source located on the platform

B. Transmission of energy from the source to the object

- Absorption and scattering of the EMR while transmission

C. Interaction of EMR with the object and subsequent reflection and emission

D. Transmission of energy from the object to the sensor

E. Recording of energy by the sensor

- Photographic or non-photographic sensors

F. Transmission of the recorded information to the ground station

G. Processing of the data into digital or hard copy image

H. Analysis of data

These stages are shown in the following figure:

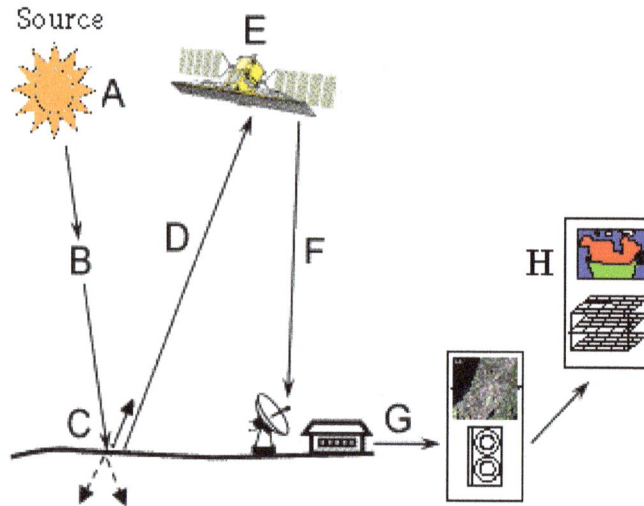

Important stages in remote sensing

Passive/ Active Remote Sensing

Depending on the source of electromagnetic energy, remote sensing can be classified as passive or active remote sensing.

In the case of passive remote sensing, source of energy is that naturally available such as the Sun. Most of the remote sensing systems work in passive mode using solar energy as the source of EMR. Solar energy reflected by the targets at specific wavelength bands are recorded using sensors onboard air-borne or space borne platforms. In order to ensure ample signal strength received at the sensor, wavelength / energy bands capable of traversing through the atmosphere, without significant loss through atmospheric interactions, are generally used in remote sensing Any object which is at a temperature above 0°K (Kelvin) emits some radiation, which is approximately proportional to the fourth power of the temperature of the object. Thus the Earth also emits some radiation since its ambient temperature is about 300°K. Passive sensors can also be used to measure the Earth's radiance but they are not very popular as the energy content is very low.

In the case of active remote sensing, energy is generated and sent from the remote sensing platform towards the targets. The energy reflected back from the targets are recorded using sensors onboard the remote sensing platform. Most of the microwave remote sensing is done through active remote sensing.

As a simple analogy, passive remote sensing is similar to taking a picture with an ordinary camera whereas active remote sensing is analogous to taking a picture with camera having built-in flash.

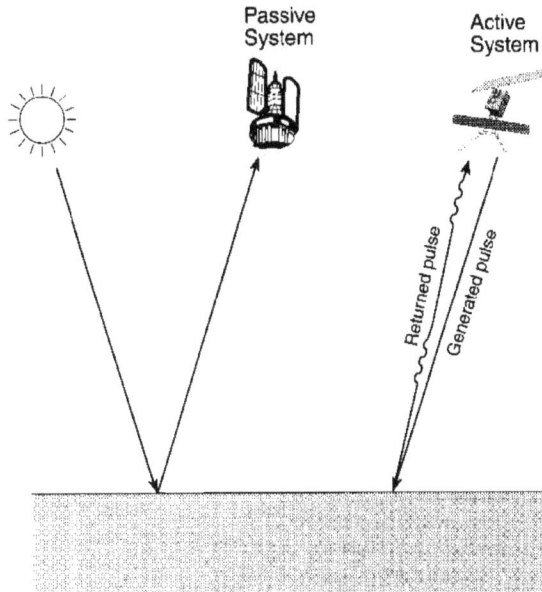

Schematic representation of passive and active remote sensing

Remote Sensing Platforms

Remote sensing platforms can be classified as follows, based on the elevation from the Earth's surface at which these platforms are placed.

- Ground level remote sensing

  o Ground level remote sensors are very close to the ground

  o They are basically used to develop and calibrate sensors for different features on the Earth's surface.

- Aerial remote sensing

  o Low altitude aerial remote sensing

  o High altitude aerial remote sensing

- Space borne remote sensing

  o Space shuttles

  o Polar orbiting satellites

  o Geo-stationary satellites

From each of these platforms, remote sensing can be done either in passive or active mode.

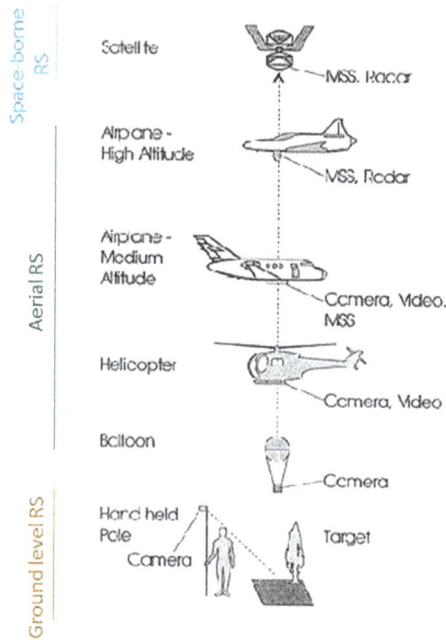

Remote sensing platforms

Airborne and Space-borne Remote Sensing

In airborne remote sensing, downward or sideward looking sensors mounted on aircrafts are used to obtain images of the earth's surface. Very high spatial resolution images (20 cm or less) can be obtained through this. However, it is not suitable to map a large area. Less coverage area and high cost per unit area of ground coverage are the major disadvantages of airborne remote sensing. While airborne remote sensing missions are mainly one-time operations, space-borne missions offer continuous monitoring of the earth features.

LiDAR, analog aerial photography, videography, thermal imagery and digital photography are commonly used in airborne remote sensing.

In space-borne remote sensing, sensors mounted on space shuttles or satellites orbiting the Earth are used. There are several remote sensing satellites (Geostationary and Polar orbiting) providing imagery for research and operational applications. While Geostationary or Geosynchronous Satellites are used for communication and meteorological purposes, polar orbiting or sun-synchronous satellites are essentially used for remote sensing. The main advantages of space-borne remote sensing are large area coverage, less cost per unit area of coverage, continuous or frequent coverage of an area of interest, automatic/ semiautomatic computerized processing and analysis. However, when compared to aerial photography, satellite imagery has a lower resolution.

Landsat satellites, Indian remote sensing (IRS) satellites, IKONOS, SPOT satellites, AQUA and TERRA of NASA and INSAT satellite series are a few examples.

Ideal Remote Sensing System

The basic components of an ideal remote sensing system include:

i.   A Uniform Energy Source which provides energy over all wavelengths, at a constant, known, high level of output

ii. A Non-interfering Atmosphere which will not modify either the energy transmitted from the source or emitted (or reflected) from the object in any manner.

iii. A Series of Unique Energy/Matter Interactions at the Earth's Surface which generate reflected and/or emitted signals that are selective with respect to wavelength and also unique to each object or earth surface feature type.

iv. A Super Sensor which is highly sensitive to all wavelengths. A super sensor would be simple, reliable, accurate, economical, and requires no power or space. This sensor yields data on the absolute brightness (or radiance) from a scene as a function of wavelength.

v.   A Real-Time Data Handling System which generates the instance radiance versus wavelength response and processes into an interpretable format in real time. The data derived is unique to a particular terrain and hence provide insight into its physical- chemical-biological state.

vi. Multiple Data Users having knowledge in their respective disciplines and also in remote sensing data acquisition and analysis techniques. The information collected will be available to them faster and at less expense. This information will aid the users in various decision making processes and also further in implementing these decisions.

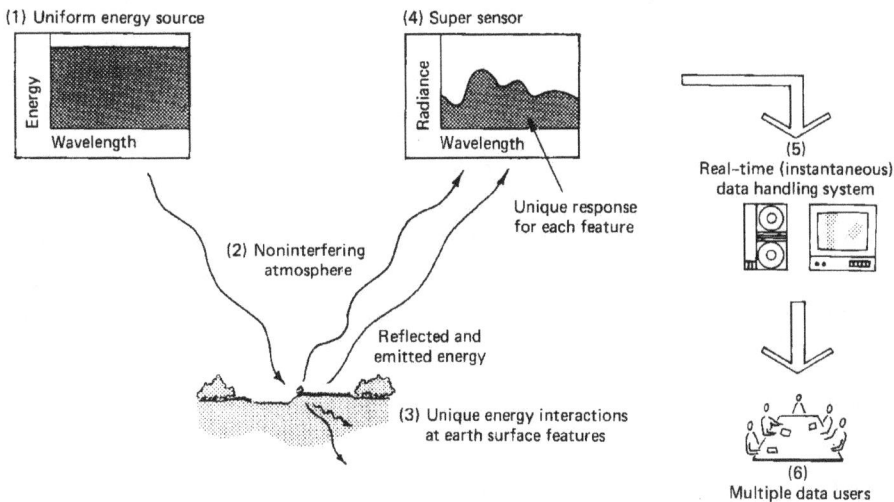

Components of an ideal remote sensing system

Characteristics of Real Remote Sensing Systems

Real remote sensing systems employed in general operation and utility have many shortcomings when compared with an ideal system explained above.

i. Energy Source: The energy sources for real systems are usually non-uniform over various wavelengths and also vary with time and space. This has major effect on the passive remote sensing systems. The spectral distribution of reflected sunlight varies both temporally and spatially. Earth surface materials also emit energy to varying degrees of efficiency. A real remote sensing system needs calibration for source characteristics.

ii. The Atmosphere: The atmosphere modifies the spectral distribution and strength of the energy received or emitted. The effect of atmospheric interaction varies with the wavelength associated, sensor used and the sensing application. Calibration is required to eliminate or compensate these atmospheric effects.

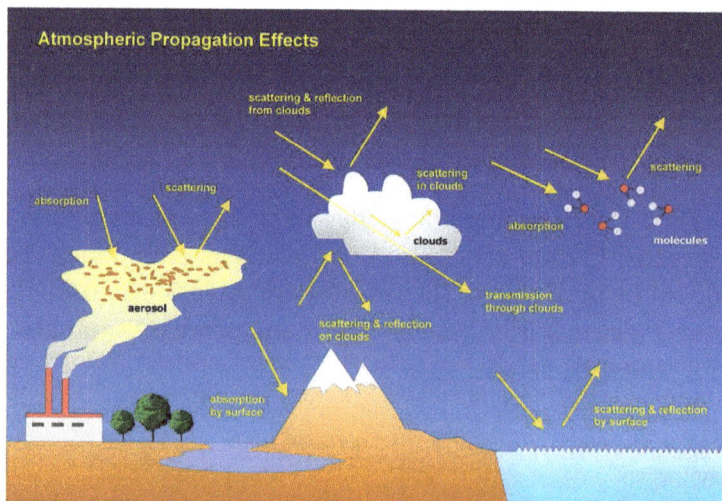

Interactions of the electromagnetic energy with the atmosphere

iii. The Energy/Matter Interactions at the Earth's Surface: Remote sensing is based on the principle that each and every material reflects or emits energy in a unique, known way. However, spectral signatures may be similar for different material types. This makes differentiation difficult. Also, the knowledge of most of the energy/matter interactions for earth surface features is either at elementary level or even completely unknown.

iv. The Sensor: Real sensors have fixed limits of spectral sensitivity i.e., they are not sensitive to all wavelengths. Also, they have limited spatial resolution (efficiency in recording spatial details). Selection of a sensor requires a trade-off between spatial resolution and spectral sensitivity. For example, while photographic systems have very good spatial resolution and poor spectral sensitivity, non-photographic systems have poor spatial resolution.

v.   The Data Handling System: Human intervention is necessary for processing sensor data; even though machines are also included in data handling. This makes the idea of real time data handling almost impossible. The amount of data generated by the sensors far exceeds the data handling capacity.

vi. The Multiple Data Users: The success of any remote sensing mission lies on the user who ultimately transforms the data into information. This is possible only if the user understands the problem thoroughly and has a wide knowledge in the data generation. The user should know how to interpret the data generated and should know how best to use them.

Advantages and Disadvantages of Remote Sensing

Advantages of remote sensing are:

a)   Provides data of large areas

b)   Provides data of very remote and inaccessible regions

c)   Able to obtain imagery of any area over a continuous period of time through which the any anthropogenic or natural changes in the landscape can be analyzed

d)   Relatively inexpensive when compared to employing a team of surveyors

e)   Easy and rapid collection of data

f)   Rapid production of maps for interpretation

Disadvantages of remote sensing are:

a)   The interpretation of imagery requires a certain skill level

b)   Needs cross verification with ground (field) survey data

c)   Data from multiple sources may create confusion

d)   Objects can be misclassified or confused

e)   Distortions may occur in an image due to the relative motion of sensor and source

EMR Spectrum

In remote sensing, some parameters of the target are measured without being in touch with it. To measure any parameters using remotely located sensors, some processes which convey those parameters to the sensor is required. A best example is the natural remote sensing by which we are able to see the objects around us and to identity their properties. We are able to see the objects around us when the solar light hits them and gets reflected and captured in our eyes. We are able to identify the properties of the objects when these signals captured in our eyes are transferred to the brain and are analysed. The whole process is analogous to the man- made remote sensing techniques.

In remote sensing techniques, electromagnetic radiations emitted / reflected by the targets are recorded at remotely located sensors and these signals are analysed to interpret the target characteristics. Characteristics of the signals recorded at the sensor depend on the characteristics of the source of radiation / energy, characteristics of the target and the atmospheric interactions.

Electromagnetic Energy

Electromagnetic (EM) energy includes all energy moving in a harmonic sinusoidal wave pattern with a velocity equal to that of light. Harmonic pattern means waves occurring at frequent intervals of time.

Electromagnetic energy has both electric and magnetic components which oscillate perpendicular to each other and also perpendicular to the direction of energy propagation as shown in the figure.

It can be detected only through its interaction with matter.

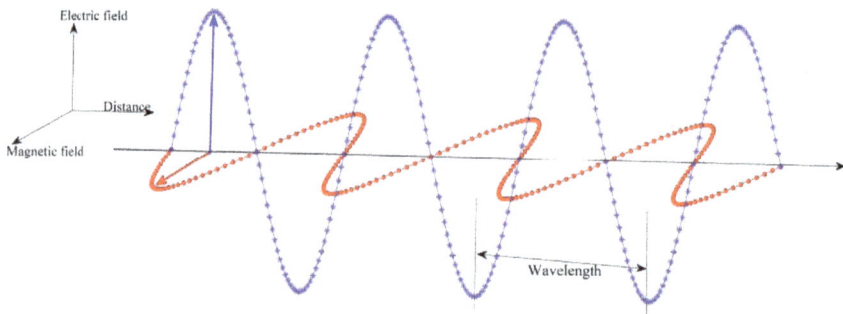

Electromagnetic wave Examples of different forms of electromagnetic energy: Light, heat etc.

EM energy can be described in terms of its velocity, wavelength and frequency.

All EM waves travel at the speed of light, c, which is approximately equal to 3×10^8 m/s.

Wavelength λ of EM wave is the distance from any point on one wave to the same position on the next wave (e.g., distance between two successive peaks). The wavelengths commonly used in remote sensing are very small. It is normally expressed in micrometers (μm). 1 μm is equal to 1×10^{-6} m.

Frequency f is the number of waves passing a fixed point per unit time. It is expressed in Hertz (Hz).

The three attributes are related by

$$c = \lambda f$$

which implies that wavelength and frequency are inversely related since c is a constant. Longer wavelengths have smaller frequency compared to shorter wavelengths.

Engineers use frequency attribute to indicate radio and radar regions. However, in remote sensing EM waves are categorized in terms of their wavelength location in the EMR spectrum.

Another important theory about the electromagnetic radiation is the particle theory, which suggests that electromagnetic radiation is composed of discrete units called photons or quanta.

Electro-Magnetic Radiation (EMR) Spectrum

Distribution of the continuum of radiant energy can be plotted as a function of wavelength (or frequency) and is known as the electromagnetic radiation (EMR) spectrum. EMR spectrum is divided into regions or intervals of different wavelengths and such regions are denoted by different names. However, there is no strict dividing line between one spectral region and its adjacent one. Different regions in EMR spectrum are indicated in the figure.

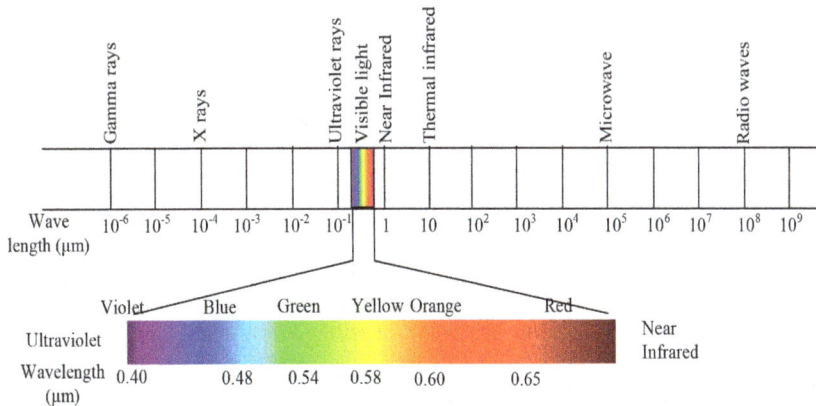

The EM spectrum ranges from gamma rays with very short wavelengths to radio waves with very long wavelengths. The EM spectrum is shown in a logarithmic scale in order to portray shorter wavelengths.

The visible region (human eye is sensitive to this region) occupies a very small region in the range between 0.4 and 0.7 µm. The approximate range of color "blue" is 0.4 – 0.5

μm, "green" is 0.5-0.6 μm and "red" is 0.6-0.7 μm. Ultraviolet (UV) region adjoins the blue end of the visible region and infrared (IR) region adjoins the red end.

The infrared (IR) region, spanning between 0.7 and 100 μm, has four subintervals of special interest for remote sensing:

(1) Reflected IR (0.7 - 3.0 μm)

(2) Film responsive subset, the photographic IR (0.7 - 0.9 μm)

(3) and (4) Thermal bands at (3 - 5 μm) and (8 - 14 μm).

Longer wavelength intervals beyond this region are referred in units ranging from 0.1 to 100 cm. The microwave region spreads across 0.1 to 100 cm, which includes all the intervals used by radar systems. The radar systems generate their own active radiation and direct it towards the targets of interest. The details of various regions and the corresponding wavelengths are given in the table.

Table: Spectrum of electromagnetic radiation.

Region	Wavelength (μm)	Remarks
Gamma rays	$< 3 \times 10^{-5}$	Not available for remote sensing. Incoming radiation is absorbed by the atmosphere
X-ray	$3 \times 10^{-5} - 3 \times 10^{-3}$	Not available for remote sensing since it is absorbed by atmosphere
Ultraviolet (UV) rays	0.03 - 0.4	Wavelengths less than 0.3 are absorbed by the ozone layer in the upper atmosphere. Wavelengths between 0.3- 0.4 μm are transmitted and termed as "Photographic UV band".
Visible	0.4 - 0.7	Detectable with film and photodetectors.
Infrared (IR)	0.7 - 100	Atmospheric windows exist which allows maximum transmission. Portion between 0.7 and 0.9 μm is called photographic IR band, since it is detectable with film. Two principal atmospheric windows exist in the thermal IR region (3 - 5 μm and 8 - 14 μm).
Microwave	$10^3 - 10^6$	Can penetrate rain, fog and clouds. Both active and passive remote sensing is possible. Radar uses wavelength in this range.
Radio	$> 10^6$	Have the longest wavelength. Used for remote sensing by some radars.

Energy in the gamma rays, X-rays and most of the UV rays are absorbed by the Earth's atmosphere and hence not used in remote sensing. Most of the remote sensing systems operate in visible, infrared (IR) and microwave regions of the spectrum. Some systems use the long wave portion of the UV spectrum also.

Energy Sources and Radiation Principles

Solar Radiation

Primary source of energy that illuminates different features on the earth surface is the Sun. Solar radiation (also called insolation) arrives at the Earth at wavelengths determined by the photosphere temperature of the sun (peaking near 5,600 °C).

Although the Sun produces electromagnetic radiation in a wide range of wavelengths, the amount of energy it produces is not uniform across all wavelengths.

The following figure shows the solar irradiance (power of electromagnetic radiation per unit area incident on a surface) distribution of the Sun. Almost 99% of the solar energy is within the wavelength range of 0.28-4.96 µm. Within this range, 43% is radiated in the visible wavelength region between 0.4-0.7 µm. The maximum energy (E) is available at 0.48 µm wave length, which is in the visible green region.

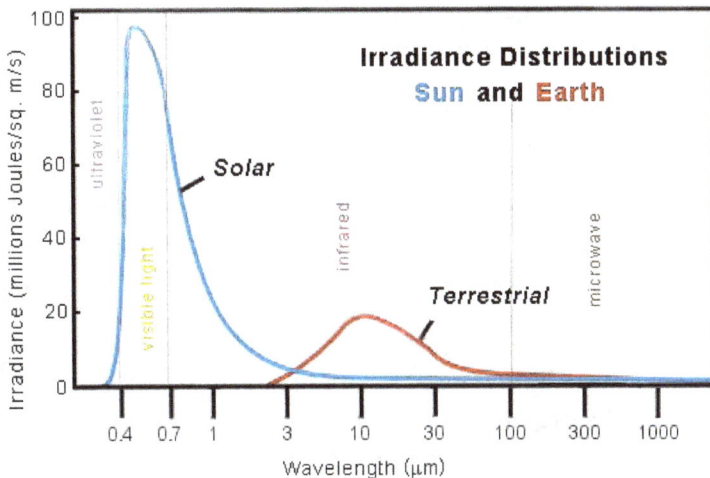

Irradiance distribution of the Sun and Earth

Sunlight

Sunlight is a portion of the electromagnetic radiation given off by the Sun, in particular infrared, visible, and ultraviolet light. On Earth, sunlight is filtered through Earth's atmosphere, and is obvious as daylight when the Sun is above the horizon. When the direct solar radiation is not blocked by clouds, it is experienced as sunshine, a combination of bright light and radiant heat. When it is blocked by clouds or reflects off other objects, it is experienced as diffused light. The World Meteorological Organization uses the term "sunshine duration" to mean the cumulative time during which an area receives direct irradiance from the Sun of at least 120 watts per square meter. Other sources indicate an "Average over the entire earth" of "164 Watts per square meter over a 24 hour day".

Sunlight shining through clouds, giving rise to crepuscular rays

The ultraviolet radiation in sunlight has both positive and negative health effects, as it is both a principal source of vitamin D_3 and a mutagen.

Sunlight takes about 8.3 minutes to reach Earth from the surface of the Sun. A photon starting at the center of the Sun and changing direction every time it encounters a charged particle would take between 10,000 and 170,000 years to get to the surface.

Sunlight is a key factor in photosynthesis, the process used by plants and other autotrophic organisms to convert light energy, normally from the Sun, into chemical energy that can be used to fuel the organisms' activities.

Measurement

Researchers may record sunlight using a sunshine recorder, pyranometer, or pyrheliometer. To calculate the amount of sunlight reaching the ground, both Earth's elliptical orbit and the attenuation by Earth's atmosphere have to be taken into account. The extraterrestrial solar illuminance (E_{ext}), corrected for the elliptical orbit by using the day number of the year (dn), is given to a good approximation by

$$E_{ext} = E_{sc} \cdot \left(1 + 0.033412 \cdot \cos\left(2\pi \frac{dn - 3}{365} \right) \right),$$

where dn=1 on January 1st; dn=32 on February 1st; dn=59 on March 1 (except on leap years, where dn=60), etc. In this formula dn−3 is used, because in modern times Earth's perihelion, the closest approach to the Sun and, therefore, the maximum E_{ext} occurs around January 3 each year. The value of 0.033412 is determined knowing that the ratio between the perihelion (0.98328989 AU) squared and the aphelion (1.01671033 AU) squared should be approximately 0.935338.

The solar illuminance constant (E_{sc}), is equal to 128×10^3 lx. The direct normal illuminance (E_{dn}), corrected for the attenuating effects of the atmosphere is given by:

$$E_{dn} = E_{ext}\, e^{-cm},$$

where c is the atmospheric extinction and m is the relative optical airmass. The atmospheric extinction brings the number of lux down to around 100 000.

The total amount of energy received at ground level from the Sun at the zenith depends on the distance to the Sun and thus on the time of year. It is about 3.3% higher than average in January and 3.3% lower in July. If the extraterrestrial solar radiation is 1367 watts per square meter (the value when the Earth–Sun distance is 1 astronomical unit), then the direct sunlight at Earth's surface when the Sun is at the zenith is about 1050 W/m², but the total amount (direct and indirect from the atmosphere) hitting the ground is around 1120 W/m². In terms of energy, sunlight at Earth's surface is around 52 to 55 percent infrared (above 700 nm), 42 to 43 percent visible (400 to 700 nm), and 3 to 5 percent ultraviolet (below 400 nm). At the top of the atmosphere, sunlight is about 30% more intense, having about 8% ultraviolet (UV), with most of the extra UV consisting of biologically damaging short-wave ultraviolet.

Direct sunlight has a luminous efficacy of about 93 lumens per watt of radiant flux. This is higher than the efficacy (of source) of most artificial lighting (including fluorescent), which means using sunlight for illumination heats up a room less than using most forms of artificial lighting.

Multiplying the figure of 1050 watts per square metre by 93 lumens per watt indicates that bright sunlight provides an illuminance of approximately 98 000 lux (lumens per square meter) on a perpendicular surface at sea level. The illumination of a horizontal surface will be considerably less than this if the Sun is not very high in the sky. Averaged over a day, the highest amount of sunlight on a horizontal surface occurs in January at the South Pole.

Dividing the irradiance of 1050 W/m² by the size of the sun's disk in steradians gives an average radiance of 15.4 MW per square metre per steradian. (However, the radiance at the centre of the sun's disk is somewhat higher than the average over the whole disk due to limb darkening.) Multiplying this by π gives an upper limit to the irradiance which can be focused on a surface using mirrors: 48.5 MW/m².

Composition and Power

The spectrum of the Sun's solar radiation is close to that of a black body with a temperature of about 5,800 K. The Sun emits EM radiation across most of the electromagnetic spectrum. Although the Sun produces gamma rays as a result of the nuclear-fusion process, internal absorption and thermalization convert these super-high-energy photons to lower-energy photons before they reach the Sun's surface and are emitted out into space. As a result, the Sun does not emit gamma rays from this process, but it does emit gamma rays from solar flares. The Sun also emits X-rays, ultraviolet, visible light, infrared, and even radio waves; the only direct signature of the nuclear process is the emission of neutrinos.

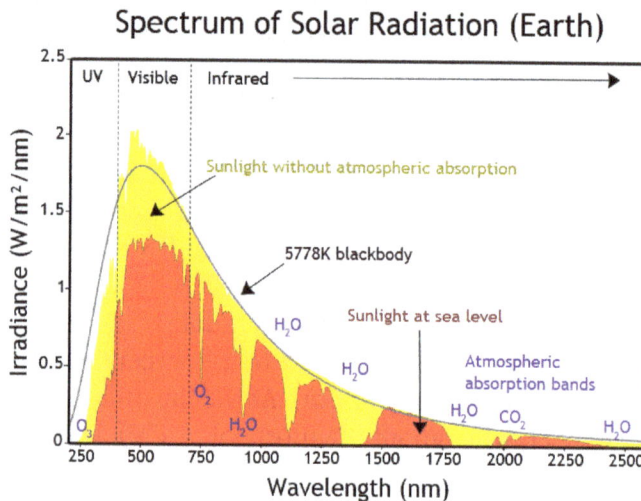

Solar irradiance spectrum above atmosphere and at surface. Extreme UV and X-rays are produced (at left of wavelength range shown) but comprise very small amounts of the Sun's total output power.

Although the solar corona is a source of extreme ultraviolet and X-ray radiation, these rays make up only a very small amount of the power output of the Sun. The spectrum of nearly all solar electromagnetic radiation striking the Earth's atmosphere spans a range of 100 nm to about 1 mm (1,000,000 nm). This band of significant radiation power can be divided into five regions in increasing order of wavelengths:

- Ultraviolet C or (UVC) range, which spans a range of 100 to 280 nm. The term *ultraviolet* refers to the fact that the radiation is at higher frequency than violet light (and, hence, also invisible to the human eye). Due to absorption by the atmosphere very little reaches Earth's surface. This spectrum of radiation has germicidal properties, as used in germicidal lamps.

- Ultraviolet B or (UVB) range spans 280 to 315 nm. It is also greatly absorbed by the Earth's atmosphere, and along with UVC causes the photochemical reaction leading to the production of the ozone layer. It directly damages DNA and causes sunburn, but is also required for vitamin D synthesis in the skin and fur of mammals.

- Ultraviolet A or (UVA) spans 315 to 400 nm. This band was once held to be less damaging to DNA, and hence is used in cosmetic artificial sun tanning (tanning booths and tanning beds) and PUVA therapy for psoriasis. However, UVA is now known to cause significant damage to DNA via indirect routes (formation of free radicals and reactive oxygen species), and can cause cancer.

- Visible range or light spans 380 to 780 nm. As the name suggests, this range is visible to the naked eye. It is also the strongest output range of the Sun's total irradiance spectrum.

- Infrared range that spans 700 nm to 1,000,000 nm (1 mm). It comprises an important part of the electromagnetic radiation that reaches Earth. Scientists divide the infrared range into three types on the basis of wavelength:

  o   Infrared-A: 700 nm to 1,400 nm

  o   Infrared-B: 1,400 nm to 3,000 nm

  o   Infrared-C: 3,000 nm to 1 mm.

Published Tables

Tables of direct solar radiation on various slopes from 0 to 60 degrees north latitude, in calories per square centimetre, issued in 1972 and published by Pacific Northwest Forest and Range Experiment Station, Forest Service, U.S. Department of Agriculture, Portland, Oregon, USA, appear on the web.

Solar Constant

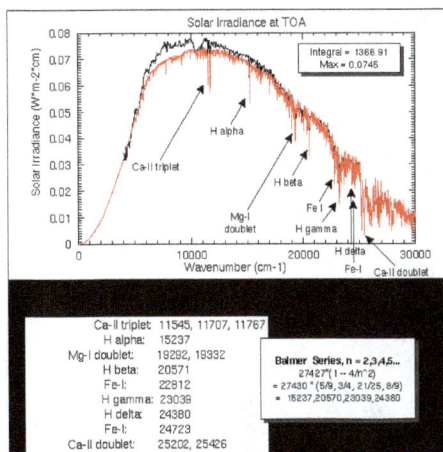

Solar irradiance spectrum at top of atmosphere, on a linear scale and plotted against wavenumber

The solar constant, a measure of flux density, is the amount of incoming solar electro-magnetic radiation per unit area that would be incident on a plane perpendicular to the rays, at a distance of one astronomical unit (AU) (roughly the mean distance from the Sun to Earth). The "solar constant" includes all types of solar radiation, not just the visible light. Its average value was thought to be approximately 1366 W/m², varying slightly with solar activity, but recent recalibrations of the relevant satellite observations indicate a value closer to 1361 W/m² is more realistic.

Total Solar Irradiance (TSI) and Spectral Solar Irradiance (SSI) upon Earth

Total solar irradiance (TSI) – the amount of solar radiation received at the top of Earth's atmosphere – has been measured since 1978 by a series of overlapping NASA and ESA satellite experiments to be 1.361 kilo watts per square meter (kW/m²). TSI observations are continuing today with the ACRIMSAT/ACRIM3, SOHO/VIRGO and SORCE/TIM satellite experiments. Variation of TSI has been discovered on many timescales including the solar magnetic cycle and many shorter periodic cycles. TSI provides the energy that drives Earth's climate, so continuation of the TSI time series database is critical to understanding the role of solar variability in climate change.

Spectral solar irradiance (SSI) – the spectral distribution of the TSI – has been monitored since 2003 by the SORCE Spectral Irradiance Monitor (SIM). It has been found that SSI at UV (ultraviolet) wavelength corresponds in a less clear, and probably more complicated fashion, with Earth's climate responses than earlier assumed, fueling broad avenues of new research in "the connection of the Sun and stratosphere, troposphere, biosphere, ocean, and Earth's climate".

Intensity in the Solar System

Sunlight on Mars is dimmer than on Earth. This photo of a Martian sunset was imaged by Mars Pathfinder.

Different bodies of the Solar System receive light of an intensity inversely proportional to the square of their distance from Sun. A rough table comparing the amount of solar radiation received by each planet in the Solar System follows (from data in):

Planet or dwarf planet	distance (AU)		Solar radiation (W/m²)	
	Perihelion	Aphelion	maximum	minimum
Mercury	0.3075	0.4667	14,446	6,272
Venus	0.7184	0.7282	2,647	2,576
Earth	0.9833	1.017	1,413	1,321
Mars	1.382	1.666	715	492
Jupiter	4.950	5.458	55.8	45.9
Saturn	9.048	10.12	16.7	13.4
Uranus	18.38	20.08	4.04	3.39
Neptune	29.77	30.44	1.54	1.47
Pluto	29.66	48.87	1.55	0.57

The actual brightness of sunlight that would be observed at the surface depends also on the presence and composition of an atmosphere. For example, Venus's thick atmosphere reflects more than 60% of the solar light it receives. The actual illumination of the surface is about 14,000 lux, comparable to that on Earth "in the daytime with overcast clouds".

Sunlight on Mars would be more or less like daylight on Earth during a slightly overcast day, and, as can be seen in the pictures taken by the rovers, there is enough diffuse sky radiation that shadows would not seem particularly dark. Thus, it would give perceptions and "feel" very much like Earth daylight. The spectrum on the surface is slightly redder than that on Earth, due to scattering by reddish dust in the Martian atmosphere.

For comparison purposes, sunlight on Saturn is slightly brighter than Earth sunlight at the average sunset or sunrise. Even on Pluto the sunlight would still be bright enough to almost match the average living room. To see sunlight as dim as full moonlight on Earth, a distance of about 500 AU (~69 light-hours) is needed; there are only a handful of objects in the Solar System known to orbit farther than such a distance, among them 90377 Sedna and (87269) 2000 OO67.

Surface Illumination

The spectrum of surface illumination depends upon solar elevation due to atmospheric effects, with the blue spectral component dominating during twilight before and after sunrise and sunset, respectively, and red dominating during sunrise and sunset. These effects are apparent in natural light photography where the principal source of illumination is sunlight as mediated by the atmosphere.

While the color of the sky is usually determined by Rayleigh scattering, an exception occurs at sunset and twilight. "Preferential absorption of sunlight by ozone over long horizon paths gives the zenith sky its blueness when the sun is near the horizon".

Spectral Composition of Sunlight at Earth's Surface

The Sun's electromagnetic radiation which is received at the Earth's surface is predominantly light that falls within the range of wavelengths to which the visual systems of the animals that inhabit Earth's surface are sensitive. The Sun may therefore be said to illuminate, which is a measure of the light within a specific sensitivity range. Many animals (including humans) have a sensitivity range of approximately 400–700 nm, and given optimal conditions the absorption and scattering by Earth's atmosphere produces illumination that approximates an equal-energy illuminant for most of this range. The useful range for color vision in humans, for example, is approximately 450–650 nm. Aside from effects that arise at sunset and sunrise, the spectral composition changes primarily in respect to how directly sunlight is able to illuminate. When illumination is indirect, Rayleigh scattering in the upper atmosphere will lead blue wavelengths to dominate. Water vapour in the lower atmosphere produces further scattering and ozone, dust and water particles will also absorb selective wavelengths.

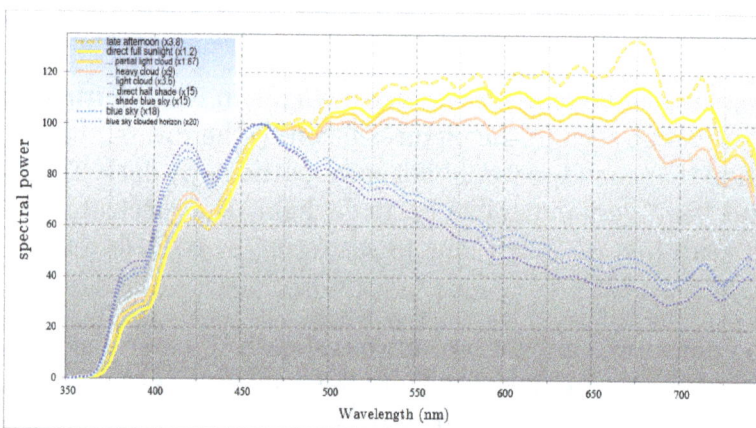

Spectrum of the visible wavelengths at approximately sea level; illumination by direct sunlight compared with direct sunlight scattered by cloud cover and with indirect sunlight by varying degrees of cloud cover. The yellow line shows the spectrum of direct illumination under optimal conditions. The other illumination conditions are scaled to show their relation to direct illumination. The units of spectral power are simply raw sensor values (with a linear response at specific wavelengths).

Variations in Solar Irradiance

Seasonal and Orbital Variation

On Earth, the solar radiation varies with the angle of the sun above the horizon, with longer sunlight duration at high latitudes during summer, varying to no sunlight at all in winter near the pertinent pole. When the direct radiation is not blocked by clouds, it is experienced as *sunshine*. The warming of the ground (and other objects) depends on the absorption of the electromagnetic radiation in the form of heat.

The amount of radiation intercepted by a planetary body varies inversely with the square of the distance between the star and the planet. Earth's orbit and obliquity change with time (over thousands of years), sometimes forming a nearly perfect circle, and at other times stretching out to an orbital eccentricity of 5% (currently 1.67%). As the orbital eccentricity changes, the average distance from the sun (the semimajor axis does not significantly vary, and so the total insolation over a year remains almost constant due to Kepler's second law,

$$\frac{2A}{r^2} dt = d\theta,$$

where A is the "areal velocity" invariant. That is, the integration over the orbital period (also invariant) is a constant.

$$\int_0 -dt = \int_0 d \ = \text{constant}.$$

If we assume the solar radiation power P as a constant over time and the solar irradiation given by the inverse-square law, we obtain also the average insolation as a constant.

But the seasonal and latitudinal distribution and intensity of solar radiation received at Earth's surface does vary. The effect of sun angle on climate results in the change in solar energy in summer and winter. For example, at latitudes of 65 degrees, this can vary by more than 25% as a result of Earth's orbital variation. Because changes in winter and summer tend to offset, the change in the annual average insolation at any given location is near zero, but the redistribution of energy between summer and winter does strongly affect the intensity of seasonal cycles. Such changes associated with the redistribution of solar energy are considered a likely cause for the coming and going of recent ice ages.

Solar Intensity Variation

Space-based observations of solar irradiance started in 1978. These measurements show that the solar constant is not constant. It varies on many time scales, including the 11-year sunspot solar cycle. When going further back in time, one has to rely on irradiance reconstructions, using sunspots for the past 400 years or cosmogenic radionuclides for going back 10,000 years. Such reconstructions have been done. These studies show that in addition to the solar irradiance variation with the solar cycle (the (Schwabe) cycle), the solar activitiy varies with longer cycles, such as the proposed 88 year (Gleisberg cycle), 208 year (DeVries cycle) and 1,000 year (Eddy cycle).

Life on Earth

The existence of nearly all life on Earth is fueled by light from the Sun. Most autotrophs, such as plants, use the energy of sunlight, combined with carbon dioxide and water, to produce simple sugars—a process known as photosynthesis. These sugars are then used as building-blocks and in other synthetic pathways that allow the organism to grow.

Heterotrophs, such as animals, use light from the Sun indirectly by consuming the products of autotrophs, either by consuming autotrophs, by consuming their products, or by consuming other heterotrophs. The sugars and other molecular components produced by the autotrophs are then broken down, releasing stored solar energy, and giving the heterotroph the energy required for survival. This process is known as cellular respiration.

In prehistory, humans began to further extend this process by putting plant and animal materials to other uses. They used animal skins for warmth, for example, or wooden weapons to hunt. These skills allowed humans to harvest more of the sunlight than was possible through glycolysis alone, and human population began to grow.

During the Neolithic Revolution, the domestication of plants and animals further increased human access to solar energy. Fields devoted to crops were enriched by inedible plant matter, providing sugars and nutrients for future harvests. Animals that had previously provided humans with only meat and tools once they were killed were now used for labour throughout their lives, fueled by grasses inedible to humans.

The more recent discoveries of coal, petroleum and natural gas are modern extensions of this trend. These fossil fuels are the remnants of ancient plant and animal matter, formed using energy from sunlight and then trapped within Earth for millions of years. Because the stored energy in these fossil fuels has accumulated over many millions of years, they have allowed modern humans to massively increase the production and consumption of primary energy. As the amount of fossil fuel is large but finite, this cannot continue indefinitely, and various theories exist as to what will follow this stage of human civilization (e.g., alternative fuels, Malthusian catastrophe, new urbanism, peak oil).

Cultural Aspects

Claude Monet: *Le déjeuner sur l'herbe*

The effect of sunlight is relevant to painting, evidenced for instance in works of Claude Monet on outdoor scenes and landscapes.

Téli verőfény ("Winter Sunshine") by László Mednyánszky

Many people find direct sunlight to be too bright for comfort, especially when reading from white paper upon which the sun is directly shining. Indeed, looking directly at the sun can cause long-term vision damage. To compensate for the brightness of sunlight, many people wear sunglasses. Cars, many helmets and caps are equipped with visors to block the sun from direct vision when the sun is at a low angle. Sunshine is often blocked from entering buildings through the use of walls, window blinds, awnings, shutters, curtains, or nearby shade trees.

In colder countries, many people prefer sunnier days and often avoid the shade. In hotter countries, the converse is true; during the midday hours, many people prefer to stay inside to remain cool. If they do go outside, they seek shade that may be provided by trees, parasols, and soon.

In Hinduism, the sun is considered to be a god, as it is the source of life and energy on earth.

Sunbathing

Sunbathing is a popular leisure activity in which a person sits or lies in direct sunshine. People often sunbathe in comfortable places where there is ample sunlight. Some common places for sunbathing include beaches, open air swimming pools, parks, gardens, and sidewalk cafes. Sunbathers typically wear limited amounts of clothing or some simply go nude. For some, an alternative to sunbathing is the use of a sunbed that generates ultraviolet light and can be used indoors regardless of weather conditions. Tanning beds have been banned in a number of states in the world.

For many people with light skin, one purpose for sunbathing is to darken one's skin color (get a sun tan), as this is considered in some cultures to be attractive, associated with outdoor activity, vacations/holidays, and health. Some people prefer naked sunbathing so that an "all-over" or "even" tan can be obtained, sometimes as part of a specific lifestyle.

For people suffering from psoriasis, sunbathing is an effective way of healing the symptoms.

Skin tanning is achieved by an increase in the dark pigment inside skin cells called melanocytes, and is an automatic response mechanism of the body to sufficient exposure to ultraviolet radiation from the sun or from artificial sunlamps. Thus, the tan gradually disappears with time, when one is no longer exposed to these sources.

Effects on Human Health

The ultraviolet radiation in sunlight has both positive and negative health effects, as it is both a principal source of vitamin D_3 and a mutagen. A dietary supplement can supply vitamin D without this mutagenic effect, but bypasses natural mechanisms that would prevent overdoses of vitamin D generated internally from sunlight. Vitamin D has a wide range of positive health effects, which include strengthening bones and possibly inhibiting the growth of some cancers. Sun exposure has also been associated with the timing of melatonin synthesis, maintenance of normal circadian rhythms, and reduced risk of seasonal affective disorder.

Long-term sunlight exposure is known to be associated with the development of skin cancer, skin aging, immune suppression, and eye diseases such as cataracts and macular degeneration. Short-term overexposure is the cause of sunburn, snow blindness, and solar retinopathy.

UV rays, and therefore sunlight and sunlamps, are the only listed carcinogens that are known to have health benefits, and a number of public health organizations state that there needs to be a balance between the risks of having too much sunlight or too little. There is a general consensus that sunburn should always be avoided.

Epidemiological data shows that people who have more exposure to the sun have less high blood pressure and cardiovascular-related mortality. While sunlight (and its UV rays) are a risk factor for skin cancer, "sun avoidance may carry more of a cost than benefit for over-all good health." A study found that there is no evidence that UV reduces lifespan in contrast to other risk factors like smoking, alcohol and high blood pressure.

Radiation from the Earth

Other than the solar radiation, the Earth and the terrestrial objects also are the sources of electromagnetic radiation. All matter at temperature above absolute zero (0°K or -273°C) emits electromagnetic radiations continuously. The amount of radiation from such objects is a function of the temperature of the object as shown below.

$$M = \sigma T^4$$

This is known as Stefan-Boltzmann law. M is the total radiant exitance from the source

(Watts / m^2), σ is the Stefan-Boltzmann constant (5.6697 x 10^{-8} Watts m^{-2}k^{-4}) and T is the absolute temperature of the emitting material in Kelvin.

Since the Earth's ambient temperature is about 300 K, it emits electromagnetic radiations, which is maximum in the wavelength region of 9.7 μm, as shown in the figure. This is considered as thermal IR radiation. This thermal IR emission from the Earth can be sensed using scanners and radiometers.

According to the Stefan-Boltzmann law, the radiant exitance increases rapidly with the temperature. However, this law is applicable for objects that behave as a blackbody.

Black-body Radiation

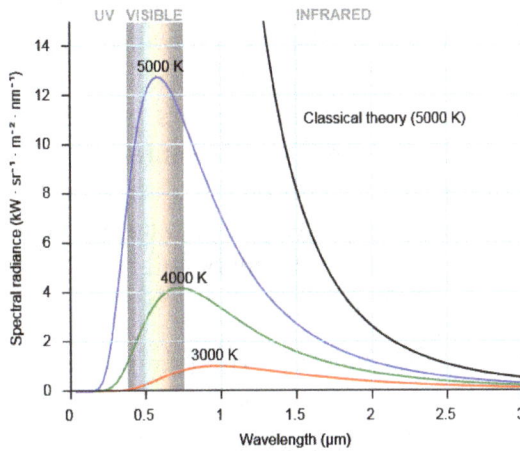

As the temperature decreases, the peak of the black-body radiation curve moves to lower intensities and longer wavelengths. The black-body radiation graph is also compared with the classical model of Rayleigh and Jeans.

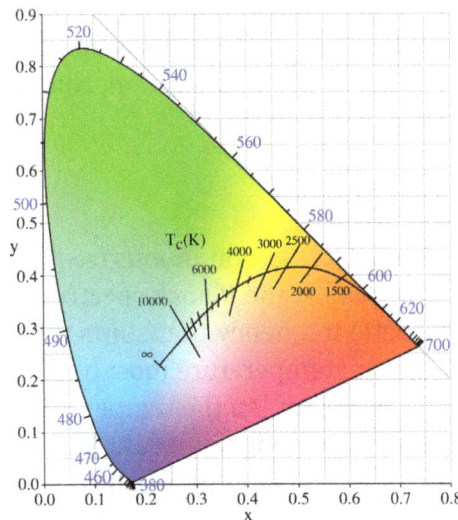

The color (chromaticity) of black-body radiation depends on the temperature of the black body; the locus of such colors, shown here in CIE 1931 x,y space, is known as the Planckian locus.

Black-body radiation is the thermal electromagnetic radiation within or surrounding a body in thermodynamic equilibrium with its environment, or emitted by a black body (an opaque and non-reflective body). It has a specific spectrum and intensity that depends only on the body's temperature, which is assumed for the sake of calculations and theory to be uniform and constant.

The thermal radiation spontaneously emitted by many ordinary objects can be approximated as black-body radiation. A perfectly insulated enclosure that is in thermal equilibrium internally contains black-body radiation and will emit it through a hole made in its wall, provided the hole is small enough to have negligible effect upon the equilibrium.

A black-body at room temperature appears black, as most of the energy it radiates is infra-red and cannot be perceived by the human eye. Because the human eye cannot perceive color at very low light intensities, a black body, viewed in the dark at the lowest just faintly visible temperature, subjectively appears grey (but only because the human eye is sensitive only to black and white at very low intensities - in reality, the frequency of the light in the visible range would still be red, although the intensity would be too low to discern as red), even though its objective physical spectrum peaks in the infrared range. When it becomes a little hotter, it appears dull red. As its temperature increases further it eventually becomes blue-white.

Although planets and stars are neither in thermal equilibrium with their surroundings nor perfect black bodies, black-body radiation is used as a first approximation for the energy they emit. Black holes are near-perfect black bodies, in the sense that they absorb all the radiation that falls on them. It has been proposed that they emit black-body radiation (called Hawking radiation), with a temperature that depends on the mass of the black hole.

The term *black body* was introduced by Gustav Kirchhoff in 1860. Black-body radiation is also called thermal radiation, *cavity radiation, complete radiation* or *temperature radiation*.

Spectrum

Black-body radiation has a characteristic, continuous frequency spectrum that depends only on the body's temperature, called the Planck spectrum or Planck's law. The spectrum is peaked at a characteristic frequency that shifts to higher frequencies with increasing temperature, and at room temperature most of the emission is in the infrared region of the electromagnetic spectrum. As the temperature increases past about 500 degrees Celsius, black bodies start to emit significant amounts of visible light. Viewed in the dark by the human eye, the first faint glow appears as a "ghostly" grey (the visible light is actually red, but low intensity light activates only the eye's grey-level sensors). With rising temperature, the glow becomes visible even when there is some background surrounding light: first as a dull red, then yellow, and eventually a "dazzling bluish-white" as the temperature rises. When the body appears white, it is emitting a substantial fraction of its energy as ultraviolet radiation. The Sun, with an effective temperature of approximately 5800 K, is an approximate black body with an emission spectrum peaked in the central, yellow-green part of the visible spectrum, but with significant power in the ultraviolet as well.

Black-body radiation provides insight into the thermodynamic equilibrium state of cavity radiation. If each Fourier mode of the equilibrium radiation in an otherwise empty cavity with perfectly reflective walls is considered as a degree of freedom capable of exchanging energy, then, according to the equipartition theorem of classical physics, there would be an equal amount of energy in each mode. Since there are an infinite number of modes this implies infinite heat capacity (infinite energy at any non-zero temperature), as well as an unphysical spectrum of emitted radiation that grows without bound with increasing frequency, a problem known as the ultraviolet catastrophe. Instead, in quantum theory the occupation numbers of the modes are quantized, cutting off the spectrum at high frequency in agreement with experimental observation and resolving the catastrophe. The study of the laws of black bodies and the failure of classical physics to describe them helped establish the foundations of quantum mechanics.

Explanation

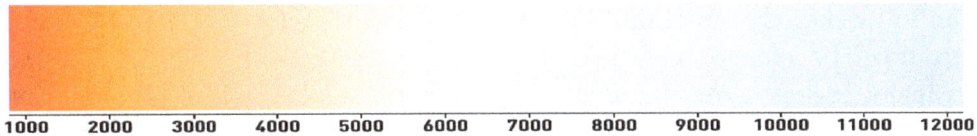

Color of a black body from 800 K to 12200 K. This range of colors approximates the range of colors of stars of different temperatures, as seen or photographed in the night sky.

All normal (baryonic) matter emits electromagnetic radiation when it has a temperature above absolute zero. The radiation represents a conversion of a body's thermal energy into electromagnetic energy, and is therefore called thermal radiation. It is a spontaneous process of radiative distribution of entropy.

Conversely all normal matter absorbs electromagnetic radiation to some degree. An object that absorbs all radiation falling on it, at all wavelengths, is called a black body. When a black body is at a uniform temperature, its emission has a characteristic frequency distribution that depends on the temperature. Its emission is called black-body radiation.

The concept of the black body is an idealization, as perfect black bodies do not exist in nature. Graphite and lamp black, with emissivities greater than 0.95, however, are good approximations to a black material. Experimentally, black-body radiation may be established best as the ultimately stable steady state equilibrium radiation in a cavity in a rigid body, at a uniform temperature, that is entirely opaque and is only partly reflective. A closed box of graphite walls at a constant temperature with a small hole on one side produces a good approximation to ideal black-body radiation emanating from the opening.

Black-body radiation has the unique absolutely stable distribution of radiative intensity that can persist in thermodynamic equilibrium in a cavity. In equilibrium, for each frequency the total intensity of radiation that is emitted and reflected from a body (that is, the net amount of radiation leaving its surface, called the *spectral radiance*) is determined solely by the equilibrium temperature, and does not depend upon the shape, material or structure of the body. For a black body (a perfect absorber) there is no reflected radiation, and so the spectral radiance is due entirely to emission. In addition, a black body is a diffuse emitter (its emission is independent of direction). Consequently, black-body radiation may be viewed as the radiation from a black body at thermal equilibrium.

Black-body radiation becomes a visible glow of light if the temperature of the object is high enough. The Draper point is the temperature at which all solids glow a dim red, about 798 K. At 1000 K, a small opening in the wall of a large uniformly heated opaque-walled cavity (let us call it an oven), viewed from outside, looks red; at 6000 K, it looks white. No matter how the oven is constructed, or of what material, as long as it is built so that almost all light entering is absorbed by its walls, it will contain a good approximation to black-body radiation. The spectrum, and therefore color, of the light that comes out will be a function of the cavity temperature alone. A graph of the amount of energy inside the oven per unit volume and per unit frequency interval plotted versus frequency, is called the *black-body curve*. Different curves are obtained by varying the temperature.

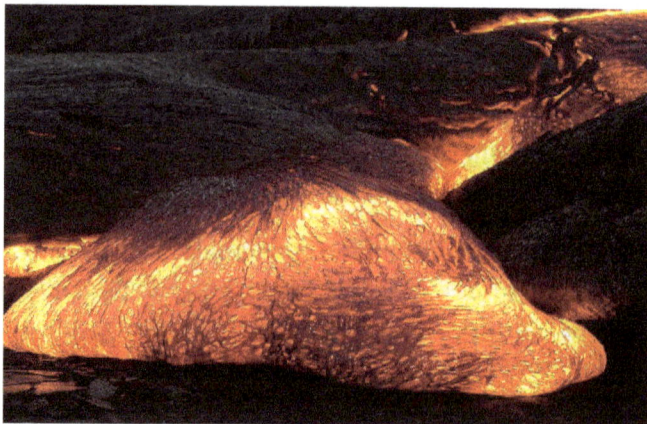

The temperature of a Pāhoehoe lava flow can be estimated by observing its color. The result agrees well with other measurements of temperatures of lava flows at about 1,000 to 1,200 °C (1,830 to 2,190 °F).

Two bodies that are at the same temperature stay in mutual thermal equilibrium, so a body at temperature T surrounded by a cloud of light at temperature T on average will emit as much light into the cloud as it absorbs, following Prevost's exchange principle, which refers to radiative equilibrium. The principle of detailed balance says that in thermodynamic equilibrium every elementary process works equally in its forward and backward sense. Prevost also showed that the emission from a body is logically determined solely by its own internal state. The causal effect of thermodynamic absorption on thermodynamic (spontaneous) emission is not direct, but is only indirect as it affects the internal state of the body. This means that at thermodynamic equilibrium the amount of every wavelength in every direction of thermal radiation emitted by a body at temperature T, black or not, is equal to the corresponding amount that the body absorbs because it is surrounded by light at temperature T.

When the body is black, the absorption is obvious: the amount of light absorbed is all the light that hits the surface. For a black body much bigger than the wavelength, the light energy absorbed at any wavelength λ per unit time is strictly proportional to the black-body curve. This means that the black-body curve is the amount of light energy emitted by a black body, which justifies the name. This is the condition for the applicability of Kirchhoff's law of thermal radiation: the black-body curve is characteristic of thermal light, which depends only on the temperature of the walls of the cavity, provided that the walls of the cavity are completely opaque and are not very reflective, and that the cavity is in thermodynamic equilibrium. When the black body is small, so that its size is comparable to the wavelength of light, the absorption is modified, because a small object is not an efficient absorber of light of long wavelength, but the principle of strict equality of emission and absorption is always upheld in a condition of thermodynamic equilibrium.

In the laboratory, black-body radiation is approximated by the radiation from a small hole in a large cavity, a hohlraum, in an entirely opaque body that is only partly reflective, that is maintained at a constant temperature. (This technique leads to the alternative term *cavity radiation*.) Any light entering the hole would have to reflect off the walls of the cavity multiple times before it escaped, in which process it is nearly certain to be absorbed. Absorption occurs regardless of the wavelength of the radiation entering (as long as it is small compared to the hole). The hole, then, is a close approximation of a theoretical black body and, if the cavity is heated, the spectrum of the hole's radiation (i.e., the amount of light emitted from the hole at each wavelength) will be continuous, and will depend only on the temperature and the fact that the walls are opaque and at least partly absorptive, but not on the particular material of which they are built nor on the material in the cavity (compare with emission spectrum).

Calculating the black-body curve was a major challenge in theoretical physics during the late nineteenth century. The problem was solved in 1901 by Max Planck in the formalism now known as Planck's law of black-body radiation. By making changes to Wien's radiation law (not to be confused with Wien's displacement law) consistent with thermodynamics and electromagnetism, he found a mathematical expression fitting the experimental data satisfactorily. Planck had to assume that the energy of the oscillators in the cavity was quantized, i.e., it existed in integer multiples of some quantity. Einstein built on this idea and proposed the quantization of electromagnetic radiation itself in 1905 to explain the photoelectric effect. These theoretical advances eventually resulted in the superseding of classical electromagnetism by quantum electrodynamics. These quanta were called photons and the black-body cavity was thought of as containing a gas of photons. In addition, it led to the development of quantum probability distributions, called Fermi–Dirac statistics and Bose–Einstein statistics, each applicable to a different class of particles, fermions and bosons.

The wavelength at which the radiation is strongest is given by Wien's displacement law, and the overall power emitted per unit area is given by the Stefan–Boltzmann law. So, as temperature increases, the glow color changes from red to yellow to white to blue. Even as the peak wavelength moves into the ultra-violet, enough radiation continues to be emitted in the blue wavelengths that the body will continue to appear blue. It will never become invisible—indeed, the radiation of visible light increases monotonically with temperature.

The radiance or observed intensity is not a function of direction. Therefore, a black body is a perfect Lambertian radiator.

Real objects never behave as full-ideal black bodies, and instead the emitted radiation at a given frequency is a fraction of what the ideal emission would be. The emissivity of a material specifies how well a real body radiates energy as compared with a black body. This emissivity depends on factors such as temperature, emission angle, and wavelength. However, it is typical in engineering to assume that a surface's spectral emissivity and absorptivity do not depend on wavelength, so that the emissivity is a constant. This is known as the *gray body* assumption.

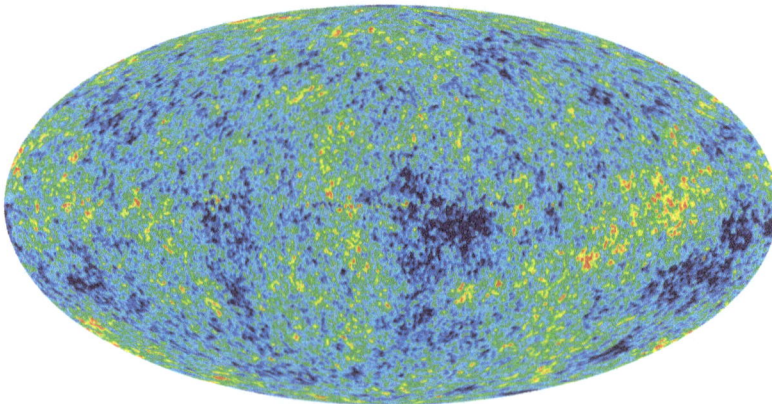

9-year WMAP image (2012) of the cosmic microwave background radiation across the universe.

With non-black surfaces, the deviations from ideal black-body behavior are determined by both the surface structure, such as roughness or granularity, and the chemical composition. On a "per wavelength" basis, real objects in states of local thermodynamic equilibrium still follow Kirchhoff's Law: emissivity equals absorptivity, so that an object that does not absorb all incident light will also emit less radiation than an ideal black body; the incomplete absorption can be due to some of the incident light being transmitted through the body or to some of it being reflected at the surface of the body.

In astronomy, objects such as stars are frequently regarded as black bodies, though this is often a poor approximation. An almost perfect black-body spectrum is exhibited by the cosmic microwave background radiation. Hawking radiation is the hypothetical black-body radiation emitted by black holes, at a temperature that depends on the mass, charge, and spin of the hole. If this prediction is correct, black holes will very gradually shrink and evaporate over time as they lose mass by the emission of photons and other particles.

A black body radiates energy at all frequencies, but its intensity rapidly tends to zero at high frequencies (short wavelengths). For example, a black body at room temperature (300 K) with one square meter of surface area will emit a photon in the visible range (390–750 nm) at an average rate of one photon every 41 seconds, meaning that for most practical purposes, such a black body does not emit in the visible range.

Equations

Planck's Law of Black-body Radiation

Planck's law states that

$$I(v,T) = \frac{2hv^3}{c^2} \frac{1}{e^{\frac{hv}{kT}} - 1},$$

where

$I(v,T)$ is the power (the energy per unit time) of frequency v radiation emitted per unit area of emitting surface in the normal direction per unit solid angle per unit frequency by a black body at temperature T, also known as spectral radiance;

h is the Planck constant;

c is the speed of light in a vacuum;

k is the Boltzmann constant;

v is the frequency of the electromagnetic radiation;

T is the absolute temperature of the body.

Wien's Displacement Law

Wien's displacement law shows how the spectrum of black-body radiation at any temperature is related to the spectrum at any other temperature. If we know the shape of the spectrum at one temperature, we can calculate the shape at any other temperature. Spectral intensity can be expressed as a function of wavelength or of frequency.

A consequence of Wien's displacement law is that the wavelength at which the intensity *per unit wavelength* of the radiation produced by a black body is at a maximum, λ_{max}, is a function only of the temperature:

$$\lambda_{max} = \frac{b}{T},$$

where the constant b, known as Wien's displacement constant, is equal to $2.8977729(17) \times 10^{-3}$ K m.

Planck's law was also stated above as a function of frequency. The intensity maximum for this is given by

$$\nu_{max} = T \times 58.8 \text{ GHz} / \text{K}.$$

Stefan–Boltzmann Law

The Stefan–Boltzmann law states that the power emitted per unit area of the surface of a black body is directly proportional to the fourth power of its absolute temperature:

$$j^{\star} = \sigma T^4,$$

where $j^{\star}$ is the total power radiated per unit area, T is the absolute temperature and $\sigma = 5.67 \times 10^{-8}$ W m^{-2} K^{-4} is the Stefan–Boltzmann constant. This follows from integrating over $I(\nu, T)$ frequency and solid angle:

$$j^{\star} = \int_0^\infty d\nu \int d\Omega \cos\theta \cdot I(\nu, T).$$

The $\cos\theta$ factor appears since we are considering the radiation in the direction normal to the surface. The solid angle integral extends over the full 2π in azimuth ϕ, and over half the domain of polar angle θ:

$$\int d\Omega \cos\theta = \int_0^{2\pi} d\phi \int_0^{\pi/2} d\theta \sin\theta \cos\theta = \pi.$$

$I(\nu, T)$ is independent of angles and passes through the solid angle integral. Inserting the formula for $I(\nu, T)$ gives

$$j^\star = \frac{2\pi(kT)^4}{c^2 h^3} \int\limits_0^\infty dx \frac{x^3}{e^x - 1},$$

where $x \equiv h\nu / kT$ is unitless. The integral over x has the value $\pi^4 >$, which gives

$$j^\star = \sigma T^4, \quad \sigma \equiv \frac{2\pi^5}{15} \frac{k^4}{c^2 h^3}.$$

Human-body Emission

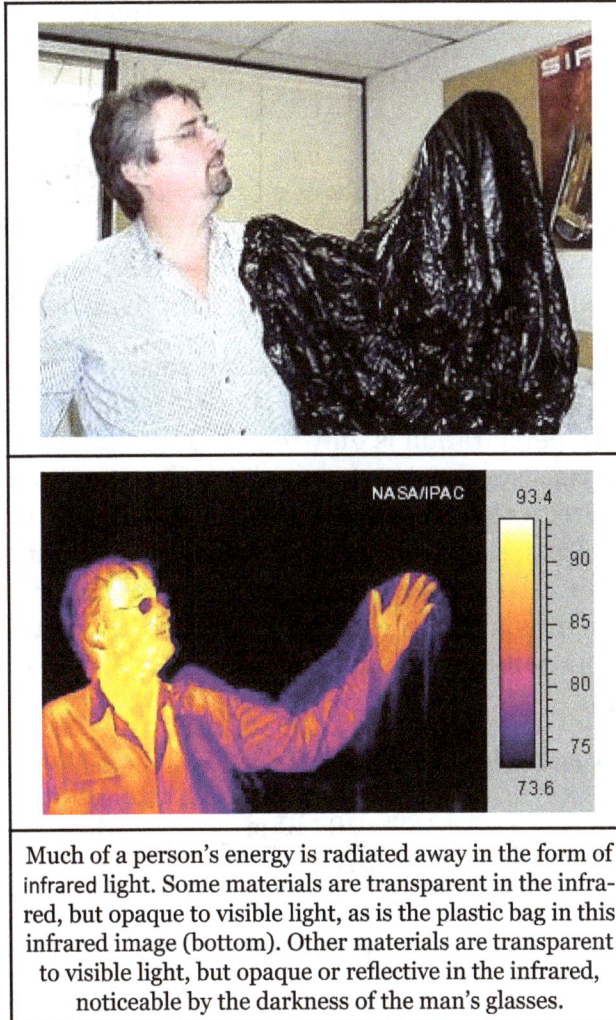

Much of a person's energy is radiated away in the form of infrared light. Some materials are transparent in the infrared, but opaque to visible light, as is the plastic bag in this infrared image (bottom). Other materials are transparent to visible light, but opaque or reflective in the infrared, noticeable by the darkness of the man's glasses.

The human body radiates energy as infrared light. The net power radiated is the difference between the power emitted and the power absorbed:

$$P_{net} = P_{emit} - P_{absorb}.$$

Applying the Stefan–Boltzmann law,

$$P_{net} = A\sigma\varepsilon\left(T^4 - T_0^4\right),$$

where A and T are the body surface area and temperature, ε is the emissivity, and T_0 is the ambient temperature.

The total surface area of an adult is about 2 m², and the mid- and far-infrared emissivity of skin and most clothing is near unity, as it is for most nonmetallic surfaces. Skin temperature is about 33 °C, but clothing reduces the surface temperature to about 28 °C when the ambient temperature is 20 °C. Hence, the net radiative heat loss is about:

$$P_{net} = 100\,\text{W}.$$

The total energy radiated in one day is about 9 MJ, or 2000 kcal (food calories). Basal metabolic rate for a 40-year-old male is about 35 kcal/(m²·h), which is equivalent to 1700 kcal per day, assuming the same 2 m² area. However, the mean metabolic rate of sedentary adults is about 50% to 70% greater than their basal rate.

There are other important thermal-loss mechanisms, including convection and evaporation. Conduction is negligible – the Nusselt number is much greater than unity. Evaporation by perspiration is only required if radiation and convection are insufficient to maintain a steady-state temperature (but evaporation from the lungs occurs regardless). Free-convection rates are comparable, albeit somewhat lower, than radiative rates. Thus, radiation accounts for about two-thirds of thermal energy loss in cool, still air. Given the approximate nature of many of the assumptions, this can only be taken as a crude estimate. Ambient air motion, causing forced convection, or evaporation reduces the relative importance of radiation as a thermal-loss mechanism.

Application of Wien's law to human-body emission results in a peak wavelength of

$$\lambda_{peak} = \frac{2.898 \times 10^{-3}\,\text{K}\cdot\text{m}}{305\,\text{K}} = 9.50\,\mu\text{m}.$$

For this reason, thermal imaging devices for human subjects are most sensitive in the 7–14 micrometer range.

Temperature Relation between a Planet and Its Star

The black-body law may be used to estimate the temperature of a planet orbiting the Sun.

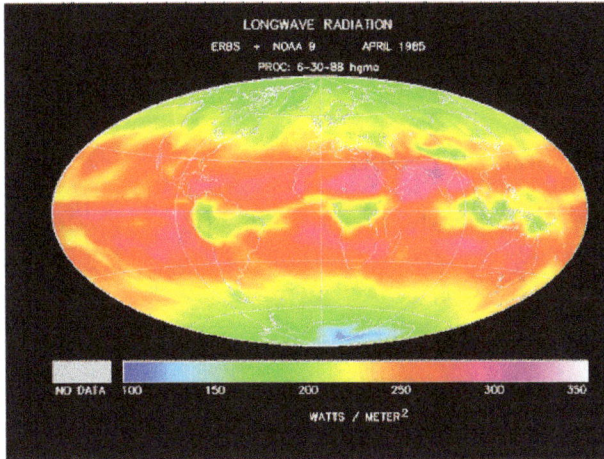

Earth's longwave thermal radiation intensity, from clouds, atmosphere and ground

The temperature of a planet depends on several factors:

- Incident radiation from its star

- Emitted radiation of the planet, e.g., Earth's infrared glow

- The albedo effect causing a fraction of light to be reflected by the planet

- The greenhouse effect for planets with an atmosphere

- Energy generated internally by a planet itself due to radioactive decay, tidal heating, and adiabatic contraction due by cooling.

The analysis only considers the Sun's heat for a planet in a Solar System.

The Stefan–Boltzmann law gives the total power (energy/second) the Sun is emitting:

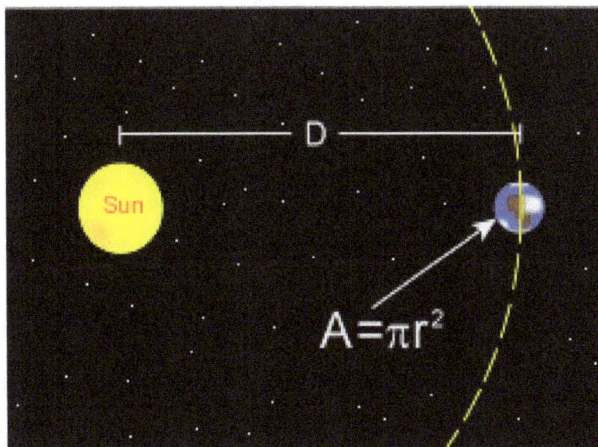

The Earth only has an absorbing area equal to a two dimensional disk, rather than the surface of a sphere.

$$P_{S\,emt} = 4\pi R_S^2 \sigma T_S^4 \qquad (1)$$

where

σ is the Stefan–Boltzmann constant,

T_S is the effective temperature of the Sun, and

R_S is the radius of the Sun.

The Sun emits that power equally in all directions. Because of this, the planet is hit with only a tiny fraction of it. The power from the Sun that strikes the planet (at the top of the atmosphere) is:

$$P_{SE} = P_{S\,emt} \left(\frac{\pi R_E^2}{4\pi D^2} \right) \qquad (2)$$

where

R_E is the radius of the planet and

D is the distance between the Sun and the planet.

Because of its high temperature, the Sun emits to a large extent in the ultraviolet and visible (UV-Vis) frequency range. In this frequency range, the planet reflects a fraction α of this energy where α is the albedo or reflectance of the planet in the UV-Vis range. In other words, the planet absorbs a fraction $1 - \alpha$ of the Sun's light, and reflects the rest. The power absorbed by the planet and its atmosphere is then:

$$P_{abs} = (1 - \alpha)P_{SE} \qquad (3)$$

Even though the planet only absorbs as a circular area πR^2, it emits equally in all directions as a sphere. If the planet were a perfect black body, it would emit according to the Stefan–Boltzmann law

$$P_{emt\,bb} = 4\pi R_E^2 \sigma T_E^4 \qquad (4)$$

where T_E is the temperature of the planet. This temperature, calculated for the case of the planet acting as a black body by setting $P_{abs} = P_{emt\,bb}$, is known as the effective temperature. The actual temperature of the planet will likely be different, depending on its surface and atmospheric properties. Ignoring the atmosphere and greenhouse effect, the planet, since it is at a much lower temperature than the Sun, emits mostly in the infrared (IR) portion of the spectrum. In this frequency range, it emits $\bar{\epsilon}$ of the radiation that a black body would emit where $\bar{\epsilon}$ is the average emissivity in the IR range. The power emitted by the planet is then:

$$P_{emt} = \bar{\epsilon} P_{emt\,bb} \qquad (5)$$

For a body in radiative exchange equilibrium with its surroundings, the rate at which it emits radiant energy is equal to the rate at which it absorbs it:

$$P_{abs} = P_{emt} \qquad (6)$$

Substituting the expressions for solar and planet power in equations 1–6 and simplifying yields the estimated temperature of the planet, ignoring greenhouse effect, T_p:

$$T_P = T_S \sqrt{\frac{R_S \sqrt{\frac{1-\alpha}{\bar{\varepsilon}}}}{2D}} \qquad (7)$$

In other words, given the assumptions made, the temperature of a planet depends only on the surface temperature of the Sun, the radius of the Sun, the distance between the planet and the Sun, the albedo and the IR emissivity of the planet.

Notice that a gray (flat spectrum) ball where $(1-\alpha) = \bar{\varepsilon}$ comes to the same temperature as a black body no matter how dark or light gray.

Virtual Temperature of Earth

Substituting the measured values for the Sun and Earth yields:

$T_S = 5778\ \text{K},$

$R_S = 6.96 \times 10^8\ \text{m},$

$D = 1.496 \times 10^{11}\ \text{m},$

$\alpha = 0.306$

With the average emissivity $\bar{\varepsilon}$ set to unity, the effective temperature of the Earth is:

$T_E = 254.356\ \text{K}$

or −18.8 °C.

This is the temperature of the Earth if it radiated as a perfect black body in the infrared, assuming an unchanging albedo and ignoring greenhouse effects (which can raise the surface temperature of a body above what it would be if it were a perfect black body in all spectrums). The Earth in fact radiates not quite as a perfect black body in the infrared which will raise the estimated temperature a few degrees above the effective temperature. If we wish to estimate what the temperature of the Earth would be if it had no atmosphere, then we could take the albedo and emissivity of the Moon as a good estimate. The albedo and emissivity of the Moon are about 0.1054 and 0.95 respectively, yielding an estimated temperature of about 1.36 °C.

Estimates of the Earth's average albedo vary in the range 0.3–0.4, resulting in different estimated effective temperatures. Estimates are often based on the solar constant (total insolation power density) rather than the temperature, size, and distance of the Sun. For example, using 0.4 for albedo, and an insolation of 1400 W m⁻², one obtains an effective temperature of about 245 K. Similarly using albedo 0.3 and solar constant of 1372 W m⁻², one obtains an effective temperature of 255 K.

Cosmology

The cosmic microwave background radiation observed today is the most perfect blackbody radiation ever observed in nature, with a temperature of about 2.7 K. It is a "snapshot" of the radiation at the time of decoupling between matter and radiation in the early universe. Prior to this time, most matter in the universe was in the form of an ionized plasma in thermal, though not full thermodynamic, equilibrium with radiation.

According to Kondepudi and Prigogine, at very high temperatures (above 10^{10} K; such temperatures existed in the very early universe), where the thermal motion separates protons and neutrons in spite of the strong nuclear forces, electron-positron pairs appear and disappear spontanteously and are in thermal equilibrium with electromagnetic radiation. These particles form a part of the black body spectrum, in addition to the electromagnetic radiation.

Doppler Effect for a Moving Black Body

The relativistic Doppler effect causes a shift in the frequency f of light originating from a source that is moving in relation to the observer, so that the wave is observed to have frequency f':

$$f' = f \frac{1 - \frac{v}{c}\cos\theta}{\sqrt{1 - v^2/c^2}},$$

where v is the velocity of the source in the observer's rest frame, θ is the angle between the velocity vector and the observer-source direction measured in the reference frame of the source, and c is the speed of light. This can be simplified for the special cases of objects moving directly towards ($\theta = \pi$) or away ($\theta = 0$) from the observer, and for speeds much less than c.

Through Planck's law the temperature spectrum of a black body is proportionally related to the frequency of light and one may substitute the temperature (T) for the frequency in this equation.

For the case of a source moving directly towards or away from the observer, this reduces to

$$T' = T\sqrt{\frac{c-v}{c+v}}.$$

Here $v > 0$ indicates a receding source, and $v < 0$ indicates an approaching source.

This is an important effect in astronomy, where the velocities of stars and galaxies can reach significant fractions of c. An example is found in the cosmic microwave background radiation, which exhibits a dipole anisotropy from the Earth's motion relative to this black-body radiation field.

History

Balfour Stewart

In 1858, Balfour Stewart described his experiments on the thermal radiative emissive and absorptive powers of polished plates of various substances, compared with the powers of lamp-black surfaces, at the same temperature. Stewart chose lamp-black surfaces as his reference because of various previous experimental findings, especially those of Pierre Prevost and of John Leslie. He wrote "Lamp-black, which absorbs all the rays that fall upon it, and therefore possesses the greatest possible absorbing power, will possess also the greatest possible radiating power." More an experimenter than a logician, Stewart failed to point out that his statement presupposed an abstract general principle, that there exist either ideally in theory or really in nature bodies or surfaces that respectively have one and the same unique universal greatest possible absorbing power, likewise for radiating power, for every wavelength and equilibrium temperature.

Stewart measured radiated power with a thermo-pile and sensitive galvanometer read with a microscope. He was concerned with selective thermal radiation, which he investigated with plates of substances that radiated and absorbed selectively for different qualities of radiation rather than maximally for all qualities of radiation. He discussed the experiments in terms of rays which could be reflected and refracted, and which obeyed the Stokes-Helmholtz reciprocity principle (though he did not use an eponym for it). He did not in this paper mention that the qualities of the rays might be described by their wavelengths, nor did he use spectrally resolving apparatus such as prisms or diffraction gratings. His work was quantitative within these constraints. He made his measurements in a room temperature environment, and quickly so as to catch his bodies in a condition near the thermal equilibrium in which they had been prepared by heating to equilibrium with boiling water. His measurements confirmed that substances that emit and absorb selectively respect the principle of selective equality of emission and absorption at thermal equilibrium.

Stewart offered a theoretical proof that this should be the case separately for every selected quality of thermal radiation, but his mathematics was not rigorously valid. He made no mention of thermodynamics in this paper, though he did refer to conservation of *vis viva*. He proposed that his measurements implied that radiation was both absorbed and emitted by particles of matter throughout depths of the media in which it propagated. He applied the Helmholtz reciprocity principle to account for the material interface processes as distinct from the processes in the interior material. He did not postulate unrealizable perfectly black surfaces. He concluded that his experiments showed that in a cavity in thermal equilibrium, the heat radiated from any part of the interior bounding surface, no matter of what material it might be composed, was the same as would have been emitted from a surface of the same shape and position that would have been composed of lamp-black. He did not state explicitly that the lamp-black-coated bodies that he used as reference must have had a unique common spectral emittance function that depended on temperature in a unique way.

Gustav Kirchhoff

In 1859, not knowing of Stewart's work, Gustav Robert Kirchhoff reported the coincidence of the wavelengths of spectrally resolved lines of absorption and of emission of visible light. Importantly for thermal physics, he also observed that bright lines or dark lines were apparent depending on the temperature difference between emitter and absorber.

Kirchhoff then went on to consider bodies that emit and absorb heat radiation, in an opaque enclosure or cavity, in equilibrium at temperature T.

Here is used a notation different from Kirchhoff's. Here, the emitting power $E(T, i)$ denotes a dimensioned quantity, the total radiation emitted by a body labeled by index i at temperature T. The total absorption ratio $a(T, i)$ of that body is dimensionless, the ratio of absorbed to incident radiation in the cavity at temperature T. (In contrast with Balfour Stewart's, Kirchhoff's definition of his absorption ratio did not refer in particular to a lamp-black surface as the source of the incident radiation.) Thus the ratio $E(T, i) / a(T, i)$ of emitting power to absorption ratio is a dimensioned quantity, with the dimensions of emitting power, because $a(T, i)$ is dimensionless. Also here the wavelength-specific emitting power of the body at temperature T is denoted by $E(\lambda, T, i)$ and the wavelength-specific absorption ratio by $a(\lambda, T, i)$. Again, the ratio $E(\lambda, T, i) / a(\lambda, T, i)$ of emitting power to absorption ratio is a dimensioned quantity, with the dimensions of emitting power.

In a second report made in 1859, Kirchhoff announced a new general principle or law for which he offered a theoretical and mathematical proof, though he did not offer quantitative measurements of radiation powers. His theoretical proof was and still is considered by some writers to be invalid. His principle, however, has endured: it was that for heat rays of the same wavelength, in equilibrium at a given temperature, the wavelength-specific ratio of emitting power to absorption ratio has one and the same common value for all bodies that emit and absorb at that wavelength. In symbols, the law stated that the wavelength-specific ratio $E(\lambda, T, i) / a(\lambda, T, i)$ has one and the same value for all bodies, that is for all values of index i. In this report there was no mention of black bodies.

In 1860, still not knowing of Stewart's measurements for selected qualities of radiation, Kirchhoff pointed out that it was long established experimentally that for total heat radiation, of unselected quality, emitted and absorbed by a body in equilibrium, the dimensioned total radiation ratio $E(T, i) / a(T, i)$, has one and the same value common to all bodies, that is, for every value of the material index i. Again without measurements of radiative powers or other new experimental data, Kirchhoff then offered a fresh theoretical proof of his new principle of the universality of the value of the wavelength-specific ratio $E(\lambda, T, i) / a(\lambda, T, i)$ at thermal equilibrium. His fresh theoretical proof was and still is considered by some writers to be invalid.

But more importantly, it relied on a new theoretical postulate of "perfectly black bodies," which is the reason why one speaks of Kirchhoff's law. Such black bodies showed complete absorption in their infinitely thin most superficial surface. They correspond to Balfour Stewart's reference bodies, with internal radiation, coated with lamp-black. They were not the more realistic perfectly black bodies later considered by Planck. Planck's black bodies radiated and absorbed only by the material in their interiors; their interfaces with contiguous media were only mathematical surfaces, capable neither of absorption nor emission, but only of reflecting and transmitting with refraction.

Kirchhoff's proof considered an arbitrary non-ideal body labeled i as well as various perfect black bodies labeled BB . It required that the bodies be kept in a cavity in thermal equilibrium at temperature T . His proof intended to show that the ratio $E(\lambda, T, i) / a(\lambda, T, i)$ was independent of the nature i of the non-ideal body, however partly transparent or partly reflective it was.

His proof first argued that for wavelength λ and at temperature T, at thermal equilibrium, all perfectly black bodies of the same size and shape have the one and the same common value of emissive power $E(\lambda, T, BB)$, with the dimensions of power. His proof noted that the dimensionless wavelength-specific absorption ratio $a(\lambda, T, BB)$ of a perfectly black body is by definition exactly 1. Then for a perfectly black body, the wavelength-specific ratio of emissive power to absorption ratio $E(\lambda, T, BB) / a(\lambda, T, BB)$ is again just $E(\lambda, T, BB)$, with the dimensions of power. Kirchhoff considered, successively, thermal equilibrium with the arbitrary non-ideal body, and with a perfectly black body of the same size and shape, in place in his cavity in equilibrium at temperature T . He argued that the flows of heat radiation must be the same in each case. Thus he argued that at thermal equilibrium the ratio $E(\lambda, T, i) / a(\lambda, T, i)$ was equal to $E(\lambda, T, BB)$, which may now be denoted $B_\lambda (\lambda, T)$, a continuous function, dependent only on λ at fixed temperature T, and an increasing function of T at fixed wavelength λ, at low temperatures vanishing for visible but not for longer wavelengths, with positive values for visible wavelengths at higher temperatures, which does not depend on the nature i of the arbitrary non-ideal body. (Geometrical factors, taken into detailed account by Kirchhoff, have been ignored in the foregoing.)

Thus Kirchhoff's law of thermal radiation can be stated: *For any material at all, radiating and absorbing in thermodynamic equilibrium at any given temperature T, for every wavelength λ, the ratio of emissive power to absorptive ratio has one universal value, which is characteristic of a perfect black body, and is an emissive power which we here represent by $B_\lambda (\lambda, T)$.* (For our notation $B_\lambda (\lambda, T)$, Kirchhoff's original notation was simply *e*.)

Kirchhoff announced that the determination of the function $B_\lambda (\lambda, T)$ was a problem of the highest importance, though he recognized that there would be experimental difficulties to be overcome. He supposed that like other functions that do not depend on the properties of individual bodies, it would be a simple function. Occasionally by historians that function $B_\lambda (\lambda, T)$ has been called "Kirchhoff's (emission, universal) function," though its precise mathematical form would not be known for another forty years, till it was discovered by Planck in 1900. The theoretical proof for Kirchhoff's universality principle was worked on and debated by various physicists over the same time, and later. Kirchhoff stated later in 1860 that his theoretical proof was better than Balfour Stewart's, and in some respects it was so. Kirchhoff's 1860 paper did not mention the second law of thermodynamics, and of course did not mention the concept of entropy which had not at that time been established. In a more considered account in a book in 1862, Kirchhoff mentioned the connection of his law with Carnot's principle, which is a form of the second law.

According to Helge Kragh, "Quantum theory owes its origin to the study of thermal radiation, in particular to the "black-body" radiation that Robert Kirchhoff had first defined in 1859–1860."

Remote Sensing using Electromagnetic Radiation

As solar energy travels through atmosphere to reach the Earth, the atmosphere absorbs or backscatters a fraction of it and transmits only the remainder. Wavelength regions, through which most of the energy is transmitted through atmosphere are referred as atmospheric windows. In the following figure, EMR spectrum is shown identifying different regions with specific names starting from visible region to microwave regions. In the microwave region, different radar bands are also shown such as κ, X, C, L and P.

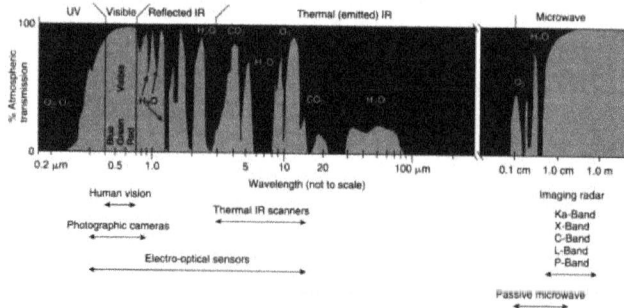

Atmospheric windows in the EMR spectrum

Energy Interaction

In many respects, remote sensing can be thought of as a reading process. Using various sensors, we remotely collect data that are analysed to obtain information about the objects, areas or phenomena being investigated. In most cases the sensors are electromagnetic sensors either air-borne or space-borne for inventorying. The sensors record the energy reflected or emitted by the target features. In remote sensing, all radiations traverse through the atmosphere for some distance to reach the sensor. As the radiation passes through the atmosphere, the gases and the particles in the atmosphere interact with them causing changes in the magnitude, wavelength, velocity, direction, and polarization.

Composition of the Atmosphere

In order to understand the interactions of the electromagnetic radiations with the atmospheric particles, basic knowledge about the composition of the atmosphere is essential.

Atmosphere is the gaseous envelop that surrounds the Earth's surface. Much of the gases are concentrated within the lower 100km of the atmosphere. Only 3×10^{-5} percent of the gases are found above 100 km (Gibbson, 2000).

Table shows the gaseous composition of the Earth's atmosphere

Table Gaseous composition of the Earth's atmosphere (from Gibbson, 2000)

Component	Percentage
Nitrogen (N_2)	78.08
Oxygen (O_2)	20.94
Argon	0.93
Carbon Dioxide (CO_2)	0.0314
Ozone (O_3)	0.00000004

Oxygen and Nitrogen are present in the ratio 1:4, and both together add to 99 percent of the total gaseous composition in the atmosphere. Ozone is present in very small quantities and is mostly concentrated in the atmosphere between 19 and 23km.

In addition to the above gases, the atmosphere also contains water vapor, methane, dust particles, pollen from vegetation, smoke particles etc. Dust particles and pollen from vegetation together form about 50 percent of the total particles present in the atmosphere. Size of these particles in the atmosphere varies from approximately 0.01μm to 100μm.

The gases and the particles present in the atmosphere cause scattering and absorption of the electromagnetic radiation passing through it.

Energy Interactions

The radiation from the energy source passes through some distance of atmosphere before being detected by the remote sensor as shown in the figure.

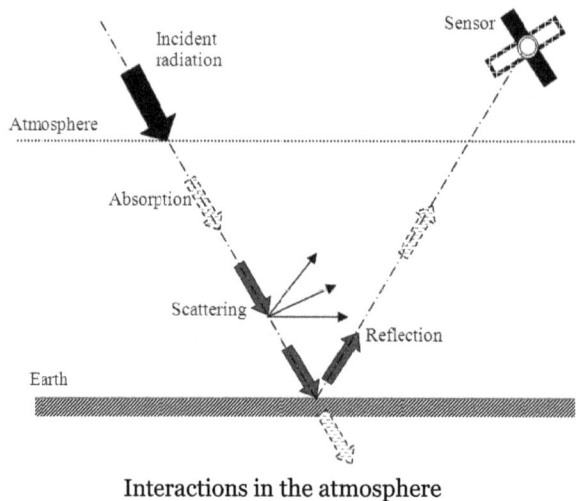

Interactions in the atmosphere

The distance travelled by the radiation through the atmosphere is called the path length. The path length varies depending on the remote sensing techniques and sources.

For example, the path length is twice the thickness of the earth's atmosphere in the case of space photography which uses sunlight as its source. For airborne thermal sensors which use emitted energy from the objects on the earth, the path length is only the length of the one way distance from the Earth's surface to the sensor, and is considerably small.

The effect of atmosphere on the radiation depends on the properties of the radiation such as magnitude and wavelength, atmospheric conditions and also the path length. Intensity and spectral composition of the incident radiation are altered by the atmospheric effects. The interaction of the electromagnetic radiation with the atmospheric particles may be a surface phenomenon (e.g., scattering) or volume phenomenon (e.g., absorption). Scattering and absorption are the main processes that alter the properties of the electromagnetic radiation in the atmosphere.

Light Scattering

Light scattering is a form of scattering in which light in the form of propagating energy is scattered. Light scattering can be thought of as the deflection of a ray from a straight path, for example by irregularities in the propagation medium, particles, or in the interface between two media. Deviations from the law of reflection due to irregularities on a surface are also usually considered to be a form of scattering. When these irregularities are considered to be random and dense enough that their individual effects average out, this kind of scattered reflection is commonly referred to as diffuse reflection.

Most objects that one sees are visible due to light scattering from their surfaces. Indeed, this is our primary mechanism of physical observation. Scattering of light depends on the wavelength or frequency of the light being scattered. Since visible light has wavelengths on the order of hundreds of nanometers, objects much smaller than this cannot be seen, even with the aid of a microscope. Colloidal particles as small as 1 μm have been observed directly in aqueous suspension.

A common form of light scattering, known as material scattering, is scattering that is attributable to the intrinsic properties of the material through which the wave is propagating. Ionospheric scattering and Rayleigh scattering are examples of material scattering. In an optical fiber, material scattering is caused by micro-inhomogeneities in the refractive indices of the materials used to fabricate the fiber, including the dopants used to modify the refractive index profile

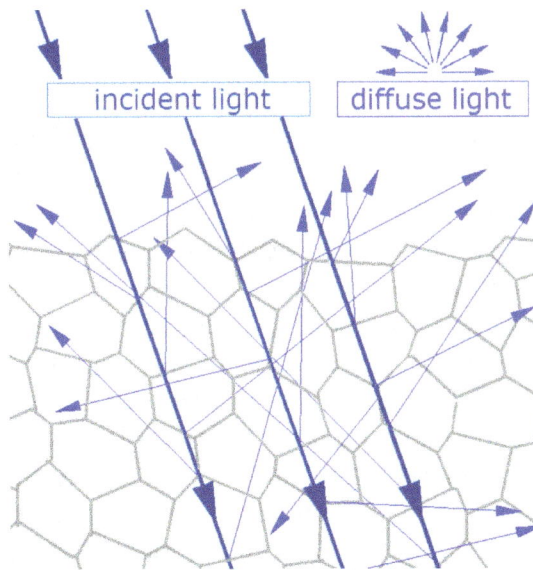

Mechanisms of diffuse reflection include surface scattering from roughness and subsurface scattering from internal irregularities such as grain boundaries in polycrystalline solids.

The transmission of various frequencies of light is essential for applications ranging from window glass to fiber optic transmission cables and infrared (IR) heat-seeking missile detection systems. Light propagating through an optical system can be attenuated by absorption, reflection and scattering.

Introduction

The interaction of light with matter can reveal important information about the structure and dynamics of the material being examined. If the scattering centers are in motion, then the scattered radiation is Doppler shifted. An analysis of the spectrum of scattered light can thus yield information regarding the motion of the scattering center. Periodicity or structural repetition in the scattering medium will cause interference in the spectrum of scattered light. Thus, a study of the scattered light intensity as a function of scattering angle gives information about the structure, spatial configuration, or morphology of the scattering medium. With regard to light scattering in liquids and solids, primary material considerations include:

- Crystalline structure: How close-packed its atoms or molecules are, and whether or not the atoms or molecules exhibit the *long-range order* evidenced in crystalline solids.

- Glassy structure: Scattering centers include fluctuations in density and/or composition.

- Microstructure: Scattering centers include internal surfaces in liquids due largely to density fluctuations, and microstructural defects in solids such as grains, grain boundaries, and microscopic pores.

In the process of light scattering, the most critical factor is the length scale of any or all of these structural features relative to the wavelength of the light being scattered.

An extensive review of light scattering in fluids has covered most of the mechanisms which contribute to the spectrum of scattered light in liquids, including density, anisotropy, and concentration fluctuations. Thus, the study of light scattering by thermally driven density fluctuations (or Brillouin scattering) has been utilized successfully for the measurement of structural relaxation and viscoelasticity in liquids, as well as phase separation, vitrification and compressibility in glasses. In addition, the introduction of dynamic light scattering and photon correlation spectroscopy has made possible the measurement of the time dependence of spatial correlations in liquids and glasses in the relaxation time gap between 10^{-6} and 10^{-2} s in addition to even shorter time scales – or faster relaxation events. It has therefore become quite clear that light scattering is an extremely useful tool for monitoring the dynamics of structural relaxation in glasses on various temporal and spatial scales and therefore provides an ideal tool for quantifying the capacity of various glass compositions for guided light wave transmission well into the far infrared portions of the electromagnetic spectrum.

- Note: Light scattering in an ideal defect-free crystalline (non-metallic) solid which provides *no scattering centers* for incoming lightwaves will be due primarily to any effects of anharmonicity within the ordered lattice. Lightwave transmission will be highly directional due to the typical anisotropy of crystalline substances, which includes their symmetry group and Bravais lattice. For example, the seven different crystalline forms of quartz silica (silicon dioxide, SiO_2) are all clear, transparent materials.

Types of Scattering

- Rayleigh scattering is the elastic scattering of light by molecules and particulate matter much smaller than the wavelength of the incident light. It occurs when light penetrates gaseous, liquid, or solid phases of matter. Rayleigh scattering intensity has a very strong dependence on the size of the particles (it is proportional to the sixth power of their diameter). It is inversely proportional to the fourth power of the wavelength of light, which means that the shorter wavelengths in visible white light (violet and blue) are scattered stronger than the longer wavelengths toward the red end of the visible spectrum. This type of scattering is therefore responsible for the blue color of the sky during the day. and the orange colors during sunrise and sunset. Rayleigh scattering is the main cause of signal loss in optical fibers.

- Mie scattering is a broad class of scattering of light by spherical particles of any diameter. The scattering intensity is generally not strongly dependent on the wavelength, but is sensitive to the particle size. Mie scattering coincides with Rayleigh scattering in the special case where the diameter of the particles is much smaller than the wavelength of the light; in this limit, however, the shape of the particles no longer matters. Mie scattering intensity for large particles is proportional to the square of the particle diameter.

- Tyndall scattering is similar to Mie scattering without the restriction to spherical geometry of the particles. It is particularly applicable to colloidal mixtures and suspensions.

- Brillouin scattering occurs from the interaction of photons with acoustic phonons in solids, which are vibrational quanta of lattice vibrations, or with elastic waves in liquids. The scattering is inelastic, meaning it is shifted in energy from the Rayleigh line frequency by an amount that corresponds to the energy of the elastic wave or phonon, and it occurs on the higher and lower energy side of the Rayleigh line, which may be associated with the creation and annihilation of a phonon. The light wave is considered to be scattered by the density maximum or amplitude of the acoustic phonon, in the same manner that X-rays are scattered by the crystal planes in a solid. In solids, the role of the crystal planes in this process is analogous to the planes of the sound waves or density fluctuations. Brillouin scattering measurements require the use of a high-contrast Fabry–Pérot interferometer to resolve the Brillouin lines from the elastic scattering, because the energy shifts are very small (< 100 cm^{-1}) and very weak in intensity. Brillouin scattering measurements yield the sound velocities in a material, which may be used to calculate the elastic constants of the sample.

- Raman scattering is another form of inelastic light scattering, but instead of scattering from acoustic phonons, as in Brillouin scattering, the light interacts with optical phonons, which are predominantly intra-molecular vibrations and rotations with energies larger than acoustic phonons. Raman scattering may therefore be used to determine chemical composition and molecular structure. Since most Raman lines are stronger than Brillouin lines, and have higher energies, standard spectrometers using scanning monochromators may be used to measure them. Raman spectrometers are standard equipment in many chemical laboratories.

Static and Dynamic Scattering

A common dichotomy in light scattering terminology is static light scattering versus dynamic light scattering. In static light scattering, the experimental variable is the time-average intensity of scattered light, whereas in dynamic light scattering it is the fluctuations in light intensity that are studied. Both techniques are typically encountered in the field of colloid and polymer characterization. They also have a broad range of other applications.

Critical Phenomena

Density fluctuations are responsible for the phenomenon of critical opalescence, which arises in the region of a continuous, or second-order, phase transition. The phenomenon is most commonly demonstrated in binary fluid mixtures, such as methanol and cyclohexane. As the critical point is approached the sizes of the gas and liquid region begin to fluctuate over increasingly large length scales. As the length scale of the density fluctuations approaches the wavelength of light, the light is scattered and causes the normally transparent fluid to appear cloudy.

References

- Wang; et al. (2005). "Modeling the Sun's Magnetic Field and Irradiance since 1713". The Astrophysical Journal. 625 (1): 522–538. Bibcode:2005ApJ...625..522W. doi:10.1086/429689

- Qiang, Fu (2003). "Radiation (Solar)" (PDF). In Holton, James R. Encyclopedia of atmospheric sciences. 5. Amsterdam: Academic Press. pp. 1859–1863. ISBN 978-0-12-227095-6. OCLC 249246073

- "Chapter 8 – Measurement of sunshine duration" (PDF). CIMO Guide. World Meteorological Organization. Retrieved 2008-12-01

- Draper, J.W. (1847). On the production of light by heat, London, Edinburgh and Dublin Philosophical Magazine and Journal of Science, series 3, 30: 345–360

- Wacker M, Holick, MF (2013). "Sunlight and Vitamin D: A global perspective for health.". Dermato-Endocrinology. 5 (1): 51–108. PMC 3897598 . PMID 24494042. doi:10.4161/derm.24494

- MacAdam, David L. (1985). Color Measurement: Theme and Variations (Second Revised ed.). Springer. pp. 33–35. ISBN 0-387-15573-2

- "NASA: The 8-minute travel time to Earth by sunlight hides a thousand-year journey that actually began in the core". NASA, sunearthday.nasa.gov. Retrieved 2012-02-12

- Willson, R. C., and A. V. Mordvinov (2003), Secular total solar irradiance trend during solar cycles 21–23, Geophys. Res. Lett., 30(5), 1199, doi:10.1029/2002GL016038 ACRIM

- Ian Morison (2008). Introduction to Astronomy and Cosmology. J Wiley & Sons. p. 48. ISBN 0-470-03333-9

- "The Multispectral Sun, from the National Earth Science Teachers Association". Windows2universe.org. 2007-04-18. Retrieved 2012-02-12

- Alessandro Fabbri; José Navarro-Salas (2005). "Chapter 1: Introduction". Modeling black hole evaporation. Imperial College Press. ISBN 1-86094-527-9

- Naylor, Mark; Kevin C. Farmer (1995). "Sun damage and prevention". Electronic Textbook of Dermatology. The Internet Dermatology Society. Retrieved 2008-06-02

- Weller, RB (2016). "Sunlight Has Cardiovascular Benefits Independently of Vitamin D.". Blood purification. 41 (1–3): 130–4. PMID 26766556. doi:10.1159/000441266

- Joseph Caniou (1999). "4.2.2: Calculation of Planck's law". Passive infrared detection: theory and applications. Springer. p. 107. ISBN 0-7923-8532-2

- "Graph of variation of seasonal and latitudinal distribution of solar radiation". Museum.state.il.us. 2007-08-30. Retrieved 2012-02-12

- J. R. Mahan (2002). Radiation heat transfer: a statistical approach (3rd ed.). Wiley-IEEE. p. 58. ISBN 978-0-471-21270-6

- "13th Report on Carcinogens: Ultraviolet-Radiation-Related Exposures" (PDF). National Toxicology Program. October 2014. Retrieved 2014-12-22

- Iribarne, J.V., Godson, W.L. (1981). Atmospheric Thermodynamics, second edition, D. Reidel Publishing, Dordrecht, ISBN 90-277-1296-4, page 227

- Lee, B. "Theoretical Prediction and Measurement of the Fabric Surface Apparent Temperature in a Simulated Man/Fabric/Environment System" (PDF). Retrieved 2007-06-24

Microwave Remote Sensing: An Overview

Microwave remote sensing is the ideal method that can be used to study atmospheric activity since microwaves can penetrate clouds, rain, light and other atmospheric phenomena. The chapter strategically encompasses and incorporates the advances in the field of remote sensing.

Passive Microwave Remote Sensing

The microwave portion of the electromagnetic spectrum involves wavelengths within a range of 1 mm to 1 m. Microwaves possess all weather operability, i.e., they are capable of penetrating clouds, haze, light rain, smoke depending on their wavelengths and hence are suitable for providing information at all times of day. This property of microwaves enables it as the best choice for atmospheric studies. Different physical principles are followed to capture images in microwave / infrared/ visible spectrum. Microwave reflections/emissions provide a different view of the earth's surface that are entirely different from those observed using visible/infrared wavelengths. In addition, unique information like sea wind/wave direction, polarization, backscattering etc are obtained which cannot be observed by sensors operating in visible/infra red regions. The only disadvantage is coarse resolution and requirement for sophisticated data analysis techniques. The portion of energy scattered to a sensor operating in microwave spectrum will depend on many factors such as dielectric constant of surface materials, type of land use/land cover, surface roughness, slopes, orientation of objects, microwave frequency, polarization, incident angle etc. Observations using microwave sensors at selected frequencies are capable of providing information regarding atmospheric structure (i.e., profiles of temperature, water vapor), liquid water content etc which make it an invaluable source for meteorological observations. The primary factors that influence the transmission of microwave signals are the wavelength and polarization of the energy pulse. These bands within the microwave spectrum were named using names originally to ensure military security during the early stages of development of radar. These traditional names have been adopted by the Institute of Electrical and Electronics Engineers and internationally by the International Telecommunication Union. These are listed below:

Table: Designation of various radar bands with their wavelengths

Band Designation	Wavelength (cm)
K_a	0.75 – 1.1
K	1.1 – 1.67
K_u	1.67 – 2.4
X	2.4 – 3.75
C	3.75 – 7.5
S	7.5 – 15
L	15 – 30
P	30 - 100

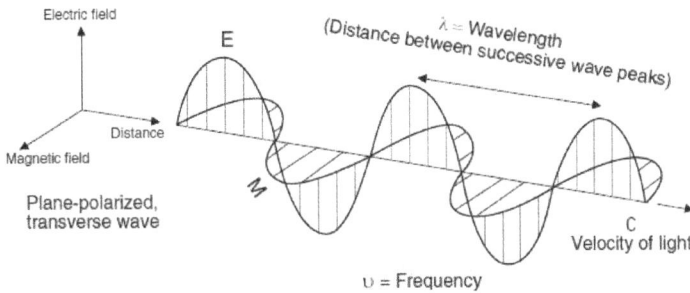

Microwave radiations are essentially characterized using intensity and polarization. Intensity is provided by the thermal emissions reaching the sensor reflected from the earth's surface after passing through the atmosphere. Intensity can be either back-scatter (for active sensor) or brightness temperature (for passive sensor). Polarization defines a group of simple waves comprising of a beam of radiation. Electromagnetic radiation is comprised of electric and magnetic radiations oscillating in mutually per-pendicular directions in such a way that at any point in space, the electric field vector of a simple electromagnetic radiation will always trace an ellipse. Polarization describes the eccentricity and orientation of this ellipse which can either be linearly polarized (i.e., in the horizontal and vertical directions) or circularly polarized.

Microwave remote sensing can be of two types: active and passive. The term "active" re-fers to sensor which transmits its own source of energy and then receives the backscat-tered energy from the surface of earth. Radar stands for radio detection and ranging. Radars like synthetic aperture radar, scatterometers, radar altimeters, ground based weather radars etc are all active instruments. Passive sensors like microwave radiom-eters receive the naturally emanating electromagnetic energy from the earth-atmo-sphere system. These sensors detect very low levels of microwave energy. The typical passive and active microwave sensors used are tabulated below:

Table: Active and Passive sensor Application

Sensor Type	Instrument	Applications
Passive Sensor	Microwave Radiometer	Sea surface Temperature (SST), Salinity, sea ice, rainfall intensity, air temperature, Near sea surface wind, ozone water vapor etc
Active Sensor	Microwave Scatterometer	Soil moisture content, water vapor, rainfall Intensity, near sea surface wind, ocean wave, biomass, sea ice, snow
	Microwave Altimeter	Sea surface topography, wind velocity, geoid, tide
	Imaging Radar	Ocean wave, topography, ice, sea surface wind, geology

Passive Remote Sensing

All natural materials emit electromagnetic radiation that is complex functions of emitting surfaces. A passive sensor operating in the microwave spectrum will usually rely on the naturally available microwave energy within their field of view instead of supplying their own source of illumination and measuring the reflected radiation like an active sensor. Passive systems utilize the electromagnetic energy that is reflected or emitted from the Earth's surface and atmosphere. Their advantages are manifold like penetration through non- precipitating clouds that aids in meteorological studies, global coverage and wide swath, highly stable instrument calibration etc. These systems suffer from large field of views (10-50 km) when compared to systems operating in the visible or infrared wavelengths. The sensors operating in passive systems use an antenna ("horn") to detect photons at microwave frequencies which are then converted to voltages in a circuit.

Principle

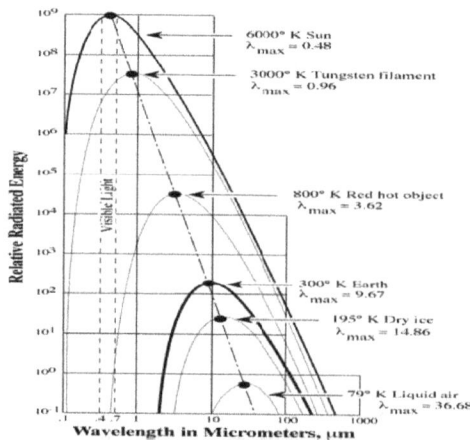

Plot of Relative radiance energy vs wavelength

The basic principle governing signal detection by passive sensors is Rayleigh Jeans's approximation of Planck's law. Conceptually, in order to understand passive microwave remote sensing, the idea of blackbody radiation theory is essential. As per thermodynamic principles, all material (gases, liquids or solids) tend to both emit as well as absorb incoherent electromagnetic energy at absolute temperature.

If B correspond to the Planck blackbody function, I be the magnitude of thermal emission and ε denote the emissivity, then thermal emission can be expressed using the relation:

$$I_\lambda = \varepsilon_\lambda * B_\lambda(T) = [\varepsilon_\lambda(2\Pi c^2 h / \lambda^5)] / (e^{hc/k\lambda T} - 1)$$

where h is Planck's constant, k is Boltzmann's constant, c the speed of light and T is thermal temperature. Once we approximate the thermal emission from Planck function using Rayleigh Jeans formula, then the microwave brightness temperature can be conveniently expressed as a linear function of physical temperature and emissivity as:

$$Tb = \varepsilon * T_{physicalTemperature}$$

where ε is a complex function of dielectric constant whose values are quite well known for gases and calm water but not so well understood for the complicated case of rough water and land surfaces.

A space borne radiometer viewing the earth senses the electromagnetic energy emanating from the land surface that reaches the top of atmosphere into the antenna after undergoing atmospheric attenuation (dampening of signal due to atmospheric components). It depends on the absorption/scattering properties of the land surface and atmosphere that tends to vary with respect to frequency and polarization. The antenna receives radiation from regions defined by the antenna pattern which is usually strongly peaked along its beam axis. Radiative transfer models are generally used to interpret the Tb received by antenna. More details can be obtained in Chandrasekhar, 1950; Wilheit et al., 1977; Volchok and Chernyak (1968) etc.

Just like thermal radiometers, microwave radiometers are non imaging devices whose output get digitally recorded on a magnetic medium. Normally, the radiometer output refers to the apparent antenna temperature. The total noise power resulting from the thermal radiation incident on the antenna, also known as "antenna temperature" is expressed as a function of the antenna gain pattern ($G(\theta,\varphi)$) and the brightness temperature distribution incident ($Tb(\theta,\varphi)$), as:

$$T_a = \frac{1}{4\Pi} \int\int_{4\Pi} Tb(\theta,\phi)G(\theta,\phi)d\Omega$$

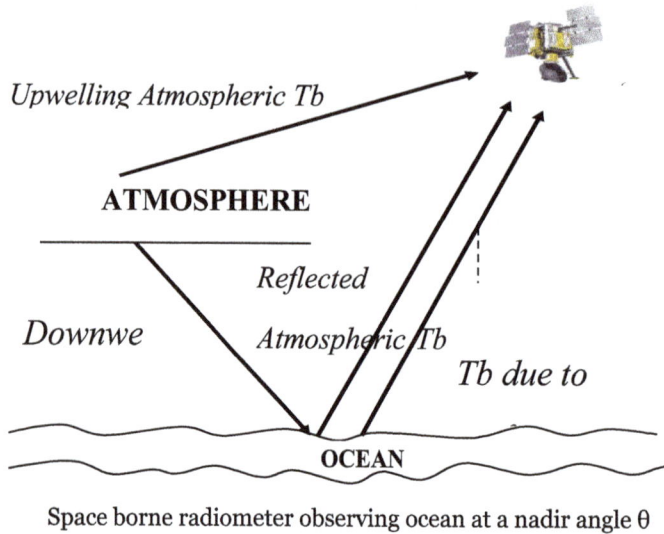

Upwelling Atmospheric Tb

ATMOSPHERE

Reflected

Downwe   *Atmospheric Tb*

Tb due to

OCEAN

Space borne radiometer observing ocean at a nadir angle θ

This system is calibrated using the temperature that a blackbody located at the antenna must reach so that the same energy is radiated as is collected from the ground scene.

Examples of Passive Microwave Radiometers

> Advanced Microwave Sounding Unit (AMSU) 1978 – present

> Scanning Multichannel Microwave Radiometer (SMMR) 1981-1987

> Special Sensor Microwave/Imager (SSM/I) 1987-present

> Tropical Rainfall Measuring Mission (TRMM) 1997-present

> Advanced Microwave Scanning Radiometer (AMSR-E)2002-present

Passive Microwave Applications

a) Soil Moisture

For passive microwave signals, in the absence of significant vegetation cover, soil moisture provides the dominant effect on the received signal. The spatial resolution of satellite borne passive microwave systems designed to study soil moisture range from 10-20 km. The low frequency ranges of 1-3 GHz is usually considered for soil moisture sensing. This is because of the reduced atmospheric attenuation and higher vegetation penetration in these wavelengths. The dielectric constant of water tends to significantly increase depending on an increase in volume fraction of water in the soil. This relationship varies with respect to different types of soil. Different soil types have different percentages of water bound on it. The bound water adhering to soil particles will exhibit rotation less freely at microwave frequencies thereby leading to a lower dielectric effect than the free water in the pore spaces. In this context, penetration depth of microwave frequencies is important as it indicates the thickness of surface layer within which moisture and temperature variations can significantly affect the emitted radiation.

b) Snow water equivalent

While viewing snow covered areas, microwave emission will comprise of contributions from snow as well as from the underlying ground. Crystals of snow will scatter part of the radiation thereby reducing the upwelling radiation measured with a radiometer. Deeper snow is indicative of greater snow crystals to scatter the upwelling microwave energy. This characteristic property is used to estimate snow mass. Usually frequencies greater than 25 GHz is utilized for detecting snow. The strength of scattering signals are proportional to the snow water equivalent. Melting water from a snow covered surface will tend to increase the microwave brightness temperature at frequencies above 30 GHz as water droplets will emit rather than scatter microwave radiation. Under these conditions, it will be difficult to extract information regarding the snow water equivalent.

c) Atmospheric water vapor

An extensive study of microwave absorption of atmospheric gases (both theoretically and experimentally) shows that, emission/absorption in gaseous atmosphere is dominated by the presence of water vapor and oxygen [*Waters*, 1976; *Ulaby et al.*, 1981]. Absorption characteristics of these gases are summarized by *Staelin* [1969], *Paris* [1971], *Derr* [1972], *Waters* [1976], *Fraser* [1975]. Microwaves undergo resonant absorption and emission at certain frequencies due to the quantum energy states of the water vapor/oxygen molecules. Within microwave spectrum, these molecules are subjected to rotational transition wherein, a molecule changes rotational energy states. This causes a peak in Tb measured by a radiometer. The magnitude of increase in Tb depends on the total number of water vapor/oxygen molecules along the propagation path through the atmosphere. An increasing altitude gets accompanied with a decrease in the number of water vapor/oxygen molecules per unit volume. This in turn reduces the bandwidth of water vapor/oxygen emission (absorption) leading to an increase in absorption at the peak of resonance. The rotational lines of water and oxygen are "pressure broadened" in the atmosphere owing to the presence of other gases; there is also a slight dependence on temperature [*Kidder and Haar*, 1995]. Water vapor has a weak absorption line at 22.235 GHz and a strong line at 183 GHz. All sensors currently used for precipitation make a measurement near this channel (SSM/I at 22.235 GHz; TMI at 21.3 GHz and AMSR at 23.8 GHz).

d) Rainfall rate

At microwave wavelengths, precipitation sized drops interact strongly with microwave radiation [*Kidder and Haar*, 1995]. Interaction of electromagnetic (EM) waves with a spherical dielectric causes scattering (redirecting) or absorption (conversion to mechanical energy) of radiation depending on size of precipitation particles [*Barrett and Martin*, 1981]. One of the earlier studies by *Mie* [1908] introduced the general mathematical solution for scattering and absorption of electromagnetic waves by a dielectric sphere of arbitrary radius. Later on, this was applied to the context of rain by *Gunn and East* [1954]. If we consider a single raindrop particle with size $<<\lambda$ of electromagnetic wave, the absorption cross section will be proportional to the volume and mass of rain drop while scattering cross section will be negligible. When cloud drop coalesce into raindrops with dimensions comparable to microwave wavelengths, absorption per unit mass increases and scattering can no longer be ignored.

Active Microwave Remote Sensing

Satellite sensors are capable of actively emitting microwaves towards the earth's surface. An active microwave system transmits electromagnetic radiation of near constant power in the form of very short pulses. These pulses will be concentrated into a narrow beam which is used for remote sensing. Active microwave systems are capable of measuring the electromagnetic waves returned from targets after they undergo reflection and atmospheric attenuation (reduction of radiation due to atmospheric particles like ice, ozone, water vapor etc). Once we know the transmission and reception times of the outgoing and incoming waves, we can easily arrive at a map showing the returned power within a three dimensional space that comprises of all the sampling volumes. This technique is generally used to track aircraft, ships or speeding automobiles. Active microwave systems can be either ground based (weather radars) or satellite based (TRMM) in nature. The most commonly used type of active microwave sensor is RADAR which is an acronym for radio detection and ranging. They have widespread applications in weather monitoring, coastal mapping, atmospheric studies, hazard mitigation studies etc. As the earth's surface represents an interface between a refractive and conducting medium, a thorough understanding of its interaction with the electromagnetic wave is essential in order to fully appreciate the detected signal. Without going into extreme detail. Some of the properties and application of synthetic aperture radars (SAR) which is an active radar is also being summarized.

Principle of Active Remote Sensing

A radar system transmits very short pulses of electromagnetic radiation concentrated into a narrow beam at predetermined radial angles. It then measures the amount of power reflected back to the radar antenna backscattered from targets within the sampling volume, as the pulse travels away from the radar. The difference between transmission and reception times of outgoing and incoming waves can be used to produce a map of returned power in three dimensional space involving all sampling volumes.

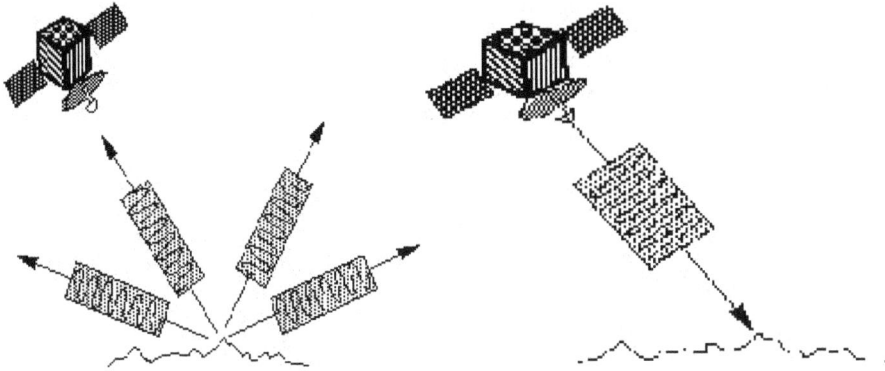

Working of radar

If P_r and P_t be the received and transmitted power, the radar equation is given as

$$P_r = 10^{-20} \frac{P'G^2\theta^2\Pi^3 h |K|^2 Z}{1024 Ln(2)\lambda^2 r^2}$$

Here G = Antenna gain [A dimensionless quantity denoting ratio of power on beam axis to power from an isotropic antenna at same point]

θ = Half power beam width (in radians)

h = Pulse width (m)

λ = Wavelength of the radar (cm)

Z = Radar reflectivity factor ($mm^6\, m^{-3}$).

If $N(D)$ denotes the raindrop size distribution in a unit volume and D denotes the diameter of the raindrop, by definition, the reflectivity factor can be expressed as proportional to the 6th moment of rain drop diameter. i.e., $Z = \int N(D)D^6\, dD$

r = Range or distance to the target (km)

K = Complex dielectric factor of the targets (dimensionless)

Radar Backscattering

The radar backscatter coefficient is given by :

$$\sigma^0 = \frac{\sigma}{A}$$

Where σ is the radar cross section and o σ denotes the radar backscatter coefficient. Radar backscatter coefficient depends on the target properties like roughness, dielectric constant and on the radar characteristics like depression angle, frequency, polarization etc. Radar backscatter is affected by dielectric constant of target (like soil). The depth of radar penetration through target like vegetation or soil will largely depend on the frequency used.

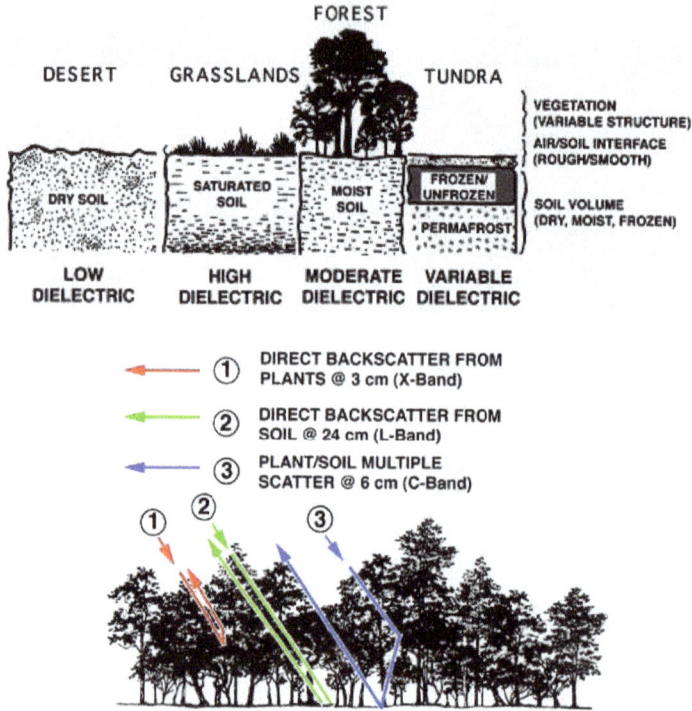

Relationship of backscatter with respect to dielectric constant and vegetation

Radar Parameters

a) Azimuth direction: Denotes the direction of aircraft or orbital track of satellite

b) Range Direction: Denotes the direction of radar illumination, usually perpendicular to the azimuth direction

c) Depression angle: Denotes the angle between horizontal plane and microwave pulse

d) Incident angle: Denotes the angle between microwave pulse and a line perpendicular to the local surface slope

e) Polarization : A simple electromagnetic wave will have electric and magnetic fields oscillating in mutually perpendicular directions. If we consider any point in space, the trajectory of the electric field vector will always trace an ellipse. Polarization is defined as the eccentricity, orientation of this ellipse and the direction along which the vector rotates.

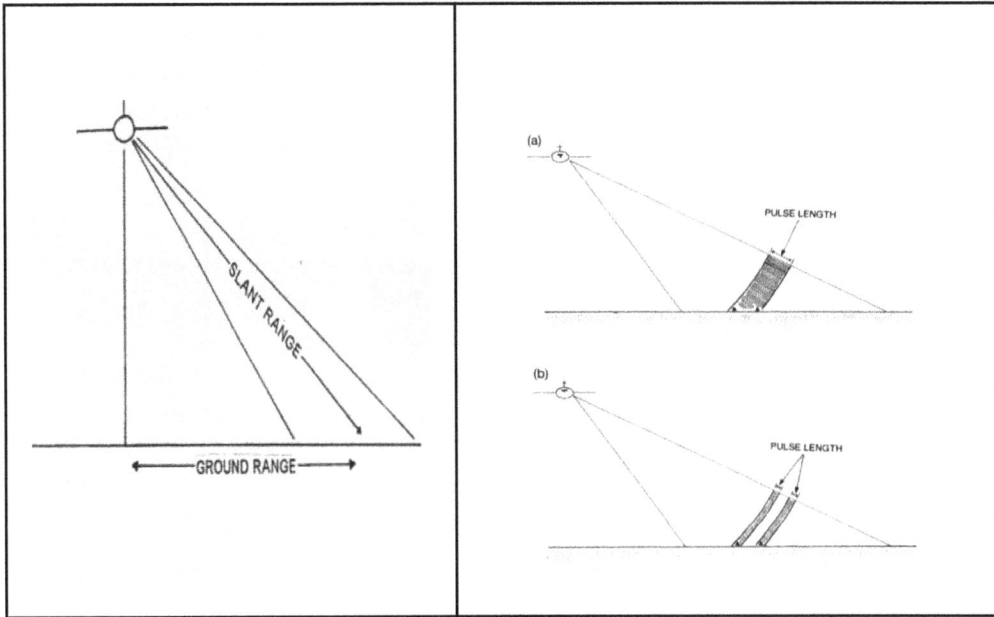

Radar parameters

Synthetic Aperture Radar (SAR)

Usually, a long radar antenna and narrower beam width results in higher azimuth resolution. Hence, precise information regarding a particular object cannot be obtained. At the same time, placing of a very large antenna in space can be very expensive. Hence, a technique is utilized which relies on satellite motion and Doppler principles in order to artificially synthesize the impression of a long antenna so that fine resolution is obtained in azimuth direction. The system which uses this technology is know as Synthetic Aperture Radar (SAR) system. As the radar moves in between two pulse transmission, it becomes possible to combine all phases of all echoes and thereby synthesize the impression of a very large antenna array.

A Synthetic Aperture Radar (SAR) is a space-borne side looking radar system which relies on the flight path to simulate an extremely large antenna or aperture electronically. SAR processors store all the radar returned signals as the platform continues to move forward. As radar measures distance to features in the slant range rather than using the true horizontal distance, radar images will be subjected to slant range distortions like foreshortening, layover etc which are discussed below. In addition, backscatter from radar can be affected by surface properties over a range of local incident angles also. For example, for incident angles of $0°$ to $30°$, topographic slope dominates the radar backscatter. For angles of $30°$ to $70°$, surface roughness dominates. Consequently, for angles > $70°$, radar shadows dominate the image.

Geometric Characteristics

Consider a tall feature tilted towards the radar (like a mountain). When the radar beam reaches the base of this tall feature before it reaches top, the radar ends up measuring the distance using slant range which will appear compressed with the length of slant feature being misrepresented. This error is called as foreshortening. Layover is the error occurring when the return signals of radar from top of target is received well before the signal from the bottom. Another effect prominent in radar images is the shadowing effect which increases with an increase in the incident angle. Unlike shadows in photography, radar shadows are completely black and are sharply defined. Following the radar shadow, a relatively weak response will be recorded from the terrain that is not oriented towards the sensor.

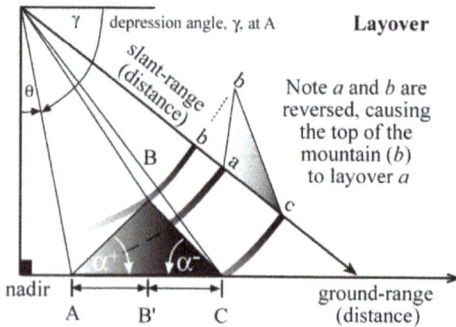

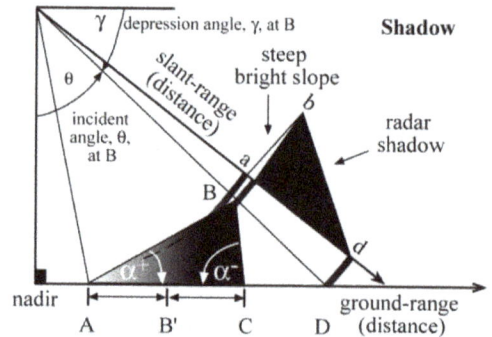

Synthetic Aperture Radar

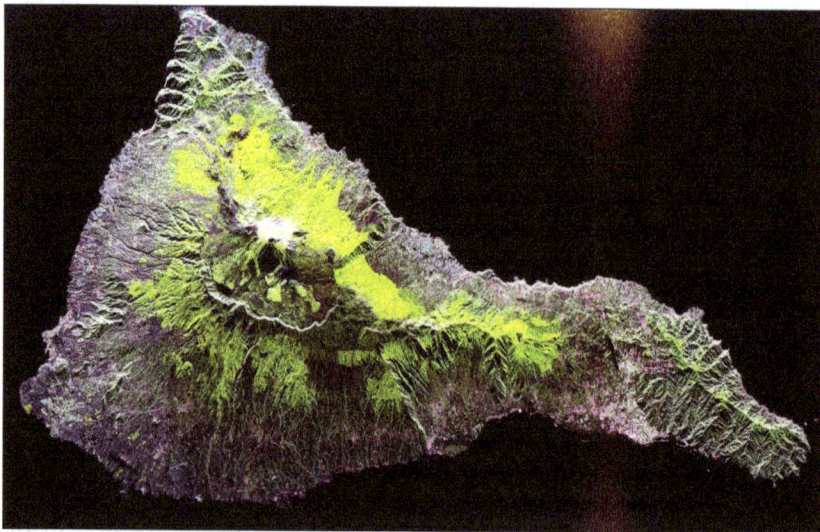

This radar image acquired by the SIR-C/X-SAR radar on board the Space Shuttle Endeavour shows the Teide volcano. The city of Santa Cruz de Tenerife is visible as the purple and white area on the lower right edge of the island. Lava flows at the summit crater appear in shades of green and brown, while vegetation zones appear as areas of purple, green and yellow on the volcano's flanks.

Synthetic aperture radar (SAR) is a form of radar that is used to create two- or 3-dimensional images of objects, such as landscapes. SAR uses the motion of the radar antenna over a target region to provide finer spatial resolution than conventional beam-scanning radars. SAR is typically mounted on a moving platform such as an aircraft or spacecraft, and has its origins in an advanced form of side-looking airborne radar (SLAR). The distance the SAR device travels over a target in the time taken for the radar pulses to return to the antenna creates the large "synthetic" antenna aperture (the "size" of the antenna). As a rule of thumb, the larger the aperture, the higher the image resolution will be, regardless of whether the aperture is physical (a large antenna) or 'synthetic' (a moving antenna) – this allows SAR to create high resolution images with comparatively small physical antennas.

To create a SAR image, successive pulses of radio waves are transmitted to "illuminate" a target scene, and the echo of each pulse is received and recorded. The pulses are transmitted and the echoes received using a single beam-forming antenna, with wavelengths of a meter down to several millimeters. As the SAR device on board the aircraft or spacecraft moves, the antenna location relative to the target changes with time. Signal processing of the successive recorded radar echoes allows the combining of the recordings from these multiple antenna positions – this process forms the 'synthetic antenna aperture', and allows the creation of higher resolution images than would otherwise be possible with a given physical antenna.

Current (2010) airborne systems provide resolutions of about 10 cm, ultra-wideband systems provide resolutions of a few millimeters, and experimental terahertz SAR has provided sub-millimeter resolution in the laboratory.

Motivation and Applications

The properties of SAR can be described as having high resolution capability which is independent of flight altitude, not being dependent on the weather as SAR can select proper frequency range. SAR also have a great day and night imaging capability considering their own illumination.

SAR images have wide applications in remote sensing and mapping of the surfaces of both the Earth and other planets. Some of the other important applications of SAR are topography, oceanography, glaciology, geology for example terrain discrimination and subsurface imaging, forestry which includes forest height, biomass, deforestation. Volcano and earthquake monitoring is a part of differential interferometry. It is also useful in environment monitoring like oil spills, flooding, urban growth, global change and military surveillance which includes strategic policy and tactical assessment. SAR can also be implemented as inverse SAR by observing a moving target over a substantial time with a stationary antenna.

Basic Principle

The surface of Venus, as imaged by the Magellan probe using SAR

A *Synthetic Aperture Radar* is an imaging radar mounted on a moving platform. Electromagnetic waves are sequentially transmitted and reflected echoes are collected, digitized and stored by the radar antenna for later processing. As transmission and reception occur at different time, they map to different positions. The well ordered combination of the received signals builds a virtual aperture that is much longer than the physical antenna length. This is why it is named "synthetic aperture", giving it the property of being an imaging radar. The range direction is parallel to flight track and perpendicular to azimuth direction, which is also known as *along-track* direction because it is in line with the position of the object within the antenna's field of view.

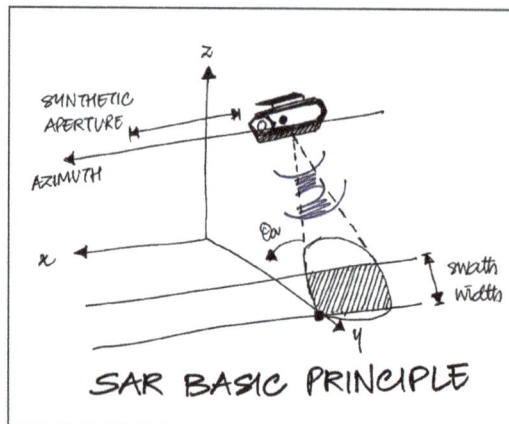

Basic principle

The 3D processing is done in two steps: the azimuth and range direction are focussed for the generation of 2D (azimuth-range) high resolution images. After which a digital elevation model (DEM) is used to measure the phase differences between complex images which is determined from different look angles to recover the height information. This height information, along with the azimuth-range coordinates provided by 2-D SAR focusing, gives the third dimension which is the elevation direction. The first step requires only standard processing algorithms, for the second step an additional pre-processing stage such as image co-registration and phase calibration is used.

In addition to this, multiple baselines can be used to extend 3D imaging to the *time dimension*. 4D and Multi-D SAR imaging allows imaging of complex scenarios, such as urban areas, and has improved performances with respect to classical interferometric techniques such as Persistent Scatterers Interferometry (PSI).

Algorithm

The SAR algorithm, as given here, applies to phased arrays generally.

A three-dimensional array (a volume) of scene elements is defined which will represent the volume of space within which targets exist. Each element of the array is a cubical voxel representing the probability (a "density") of a reflective surface being at that location in space. (Note that two-dimensional SARs are also possible—showing only a top-down view of the target area).

Initially, the SAR algorithm gives each voxel a density of zero.

Then, for each captured waveform, the entire volume is iterated. For a given waveform and voxel, the distance from the position represented by that voxel to the antenna(e) used to capture that waveform is calculated. That distance represents a time delay into the waveform. The sample value at that position in the waveform is then added to the voxel's density value. This represents a possible echo from a target at that position. Note that there are several optional approaches here, depending on the precision of the waveform timing, among other things. For example, if phase cannot be accurately known, then only the envelope magnitude (with the help of a Hilbert transform) of the waveform sample might be added to the voxel. If polarization and phase are known in the waveform, and are accurate enough, then these values might be added to a more complex voxel that holds such measurements separately.

After all waveforms have been iterated over all voxels, the basic SAR processing is complete.

What remains, in the simplest approach, is to decide what voxel density value represents a solid object. Voxels whose density is below that threshold are ignored. Note that the threshold level chosen must at least be higher than the peak energy of any single wave—otherwise that wave peak would appear as a sphere (or ellipse, in the case of multistatic operation) of false "density" across the entire volume. Thus to detect a point on a target, there must be at least two different antenna echoes from that point. Consequently, there is a need for large numbers of antenna positions to properly characterize a target.

The voxels that passed the threshold criteria are visualized in 2D or 3D. Optionally, added visual quality can sometimes be had by use of a surface detection algorithm like marching cubes.

Existing Spectral Estimation Approaches

Synthetic Aperture Radar determines the 3-D reflectivity from measured SAR data. It is basically a spectrum estimation, because for a specific cell of an image, the complex valued SAR measurements of the SAR image stack are sampled version of the Fourier transform of reflectivity in elevation direction, but the Fourier transform is irregular. Thus the spectral estimation techniques are used to improve the resolution, reduce speckle compared to what we get in conventional Fourier transform SAR imaging techniques.

Non-parametric methods

FFT

FFT is one such method which is used in majority of the spectral estimation algorithms and there are many fast algorithms for computing the multidimensional discrete Fourier transform. Computational *Kronecker-core array algebra* is a popular algorithm used as new variant of FFT algorithms for the processing in multidimensional synthetic aperture radar (SAR) systems. This algorithm uses a study of theoretical properties of input/output data indexing sets and groups of permutations.

A branch of finite multi-dimensional linear algebra, is used to identify similarities and differences among various FFT algorithm variants and also to create new variants. Each multidimensional DFT computation is expressed in matrix form. The multidimensional DFT matrix, in turn, is disintegrated into a set of factors, called functional primitives, which are individually identified with an underlying software/hardware computational design.

The FFT implementation is essentially a realization of the mapping of the mathematical framework, through generation of the variants and executing matrix operations. The performance of this implementation may vary from machine to machine and the objective is to identify on which machine it performs best.

Advantages

- Additive group theoretic properties of multidimensional input/output indexing sets are used for the mathematical formulations, therefore, it is easier to identify mapping between computing structures and mathematical expressions and thus better than conventional methods.

- The language of CKA algebra helps the application developer in understanding which are the more computational efficient FFT variants and thus reducing the computational effort and improve their implementation time.

Disadvantages

- FFT cannot separate sinusoids closer in frequency. Also if the periodicity of the data does not match FFT, edge effects are seen.

Capon Method

The Capon spectral method, also called the minimum variance method, is a multidimensional array-processing technique. It is a nonparametric covariance based method which has adaptive matched- filterbank approach and follows two main steps.

a) Passing the data through a 2-D bandpass filter with varying center frequencies (ω_1, ω_2).

b) Estimating the power at $(\omega_1\ \omega_2)$ for all $\omega_1 \in [0, 2\pi)$, $\omega_2 \in [0, 2\pi)$ of interest from the filtered data.

The adaptive Capon bandpass filter is designed to minimize the power of the filter output, as well as pass the frequencies (ω_1, ω_2) without any attenuation, i.e., to satisfy, for each (ω_1, ω_2)

$$min_h\ h^*_{\omega_1,\omega_2} R h_{\omega_1,\omega_2}\ \text{subject to}\ h^*_{\omega_1,\omega_2} a_{\omega_1,\omega_2} = 1$$

where R is the covariance matrix , $h\ \omega_1, \omega_2$ is the complex conjugate transpose of the impulse response of the FIR filter, a_{ω_1,ω_2} is the 2-D Fourier vector, defined as $a_{\omega_1,\omega_2} \triangleq a_{\omega_1} \otimes a_{\omega_2}$, $\otimes$ denotes Kronecker product.

Therefore, it passes a 2-D sinusoid at a given frequency without distortion while minimizing the variance of the noise of the resulting image. The purpose is to compute the spectral estimate efficiently.

Spectral estimate is given as :

$$S_{\omega_1,\omega_2} = \frac{1}{(a^*_{\omega_1,\omega_2} R^{-1} a_{\omega_1,\omega_2})}$$

where R is the covariance matrix and $a^*_{\omega_1,\omega_2}$ is the 2D complex conjugate transpose of the Fourier vector. The computation of this equation over all frequencies is time consuming. It is seen that the forward-backward Capon estimator yields better estimation than the forward-only classical capon approach. The main reason behind this is that while the forward-backward Capon uses both the forward and backward data vectors to obtain the estimate of the covariance matrix, the forward- only Capon uses only the forward data vectors to estimate the covariance matrix.

Advantages

- Capon can yield more accurate spectral estimates with much lower sidelobes and narrower spectral peaks than the fast Fourier transform (FFT) method.

- Capon method can provide much better resolution.

Disadvantages

- Implementation requires computation of two intensive task: inversion of the covariance matrix R and also multiply it with the a_{ω_1,ω_2} matrix which has to be done for each point (ω_1,ω_2).

APES Method

The APES (Amplitude and Phase Estimation) method is also a matched filter bank method which assumes that the phase history data is a sum of 2-D sinusoids in noise.

APES spectral estimator has 2-step filtering interpretation:

a) Passing data through a bank of FIR bandpass filters with varying center frequency, .

b) Obtaining the spectrum estimate for $\omega \in [0,2\pi)$ from the filtered data.

Empirically, the APES method results in wider spectral peaks than the Capon method, but more accurate spectral estimates for amplitude in SAR. In the Capon method, although the spectral peaks are narrower than the APES, the sidelobes are higher than that for the APES. As a result, the estimate for the amplitude is expected to be less accurate for the Capon method than for the APES method. The APES method requires about 1.5 times more computation than the Capon method.

Advantages

- Filtering reduces the number of available samples, but when it is designed tactically, the increase in signal-to-noise ratio (SNR) in the filtered data will compensate this reduction, and the amplitude of a sinusoidal component with frequency ω can be estimated more accurately from the filtered data than from the original signal.

Disadvantages

- The autocovariance matrix is much larger in 2-D than in 1-D, therefore it is limited by memory available.

Parametric Subspace Decomposition Methods

Eigenvector Method

This subspace decomposition method separates the eigenvectors of the autocovariance matrix into those corresponding to signals and to clutter. The amplitude of the image at a point (ω_x, ω_y) is given by:

$$\hat{\phi}_{EV}(\omega_x, \omega_y) = \frac{1}{W^H(\omega_x, \omega_y)\left(\sum_{clutter} \frac{1}{\lambda_i} v_i v_i^H\right)W(\omega_x, \omega_y)}$$

where $\hat{\phi}_{EV}$ is the amplitude of the image at a point (ω_x, ω_y), v_i is the coherency matrix and v_i^H is the Hermitian of the coherency matrix, $\frac{1}{\lambda_i}$ is the inverse of the eigenvalues of the clutter subspace, $W(\omega_x, \omega_y)$ are vectors defined as

$$W(\omega_x, \omega_y) = [1 \exp(-j\omega_x)....\exp(-j(M-1)\omega_x] \otimes [1 \exp(-j\omega_y)....\exp(-j(M-1)\omega_y]$$

where $\otimes$ denotes the Kronecker product of the two vectors.

Advantages

- Shows features of image more accurately.

Disadvantages

- High computational complexity.

MUSIC Method

MUSIC detects frequencies in a signal by performing an eigen decomposition on the covariance matrix of a data vector of the samples obtained from the samples of the received signal. When all of the eigenvectors are included in the clutter subspace (model order = 0) the EV method becomes identical to the Capon method. Thus the determination of model order is critical to operation of the EV method. The eigenvalue of the R matrix decides whether its corresponding eigenvector corresponds to the clutter or to the signal subspace.

The MUSIC method is considered to be a poor performer in SAR applications. This method uses a constant instead of the clutter subspace.

In this method, the denominator is equated to zero when a sinusoidal signal corresponding to a point in the SAR image is in alignment to one of the signal subspace eigenvectors which is the peak in image estimate. Thus this method does not accurately represent the scattering intensity at each point, but show the particular points of the image.

Advantages

- MUSIC whitens or equalizes, the clutter eigenvalues.

Disadvantages

- Resolution loss due to the averaging operation.

Backprojection Algorithm

Backprojection Algorithm has two methods: *Time-domain Backprojection* and *Frequency-domain Backprojection*. The time-domain Backprojection has more advantages over frequency-domain and thus, is more preferred. The time-domain Backprojection forms images or spectrums by matching the data acquired from the radar and as per what it expects to receive. It can be considered as an ideal matched-filter for Synthetic Aperture Radar. There is no need of having a different motion compensation step due to its quality of handling non-ideal motion/sampling. It can also be used for various imaging geometries.

Advantages

- *It is invariant to the imaging mode*: which means, that it uses the same algorithm irrespective of the imaging mode present, whereas, frequency domain methods require changes depending on the mode and geometry.

- Ambiguous azimuth aliasing usually occurs when the Nyquist spatial sampling requirements are exceeded by frequencies. Unambiguous aliasing occurs in squinted geometries where the signal bandwidth does not exceed the sampling limits, but has undergone "spectral wrapping." Backprojection Algorithm does not get affected by any such kind of aliasing effects.

- *It matches the space/time filter:* uses the information about the imaging geometry, to produce a pixel-by-pixel varying matched filter to approximate the expected return signal. This usually yields antenna gain compensation.

- With reference to the previous advantage, the back projection algorithm compensates for the motion. This becomes an advantage at areas having low altitudes.

Disadvantages

- The computational expense is more for Backprojection algorithm as compared to other frequency domain methods.

- It requires very precise knowledge of imaging geometry.

Application: Geosynchronous Orbit Synthetic Aperture Radar (GEO-SAR)

In GEO-SAR, to focus specially on the relative moving track, the backprojection algorithm works very well. It uses the concept of Azimuth Processing in the time domain. For the satellite-ground geometry, GEO-SAR plays a significant role.

The procedure of this concept is elaborated as follows.

1. The raw data acquired is segmented or drawn into sub-apertures for simplification of speedy conduction of procedure.

2. The range of the data is then compressed, using the concept of "Matched Filtering" for every segment/sub-aperture created. It is given by- $s(t,\tau) = \exp\{-j \cdot \left(\frac{4\pi}{\lambda}\right) \cdot R(t)\} \cdot sinc\left(\tau - \left(\frac{2 \cdot R(t)}{c}\right)\right)$ where τ is the range time, t is the azimuthal time, λ is the wavelength, c is the speed of light.

3. Accuracy in the "Range Migration Curve" is achieved by range interpolation.

4. The pixel locations of the ground in the image is dependent on the satellite–ground geometry model. Grid-division is now done as per the azimuth time.

5. Calculations for the "slant range" (range between the antenna's phase center and the point on the ground) are done for every azimuth time using coordinate transformations.

6. Azimuth Compression is done after the previous step.

7. Step 5 and 6 are repeated for every pixel, to cover every pixel, and conduct the procedure on every sub-aperture.

8. Lastly, all the sub-apertures of the image created throughout, are superimposed onto each other and the ultimate HD image is generated.

Comparison between the Algorithms

Capon and APES can yield more accurate spectral estimates with much lower sidelobes and more narrow spectral peaks than the fast Fourier transform (FFT) method, which is also a special case of the FIR filtering approaches. It is seen that although the APES algorithm gives slightly wider spectral peaks than the Capon method, the former yields more accurate overall spectral estimates than the latter and the FFT method.

FFT method is fast and simple but have larger sidelobes. Capon has high resolution but high computational complexity. EV also has high resolution and high computational complexity. APES has higher resolution, faster than capon and EV but high computational complexity.

MUSIC method is not generally suitable for SAR imaging, as whitening the clutter eigenvalues destroys the spatial inhomogeneities associated with terrain clutter or other diffuse scattering in SAR imagery. But it offers higher frequency resolution in the resulting power spectral density (PSD) than the fast Fourier transform (FFT)-based methods.

Backprojection Algorithm is computationally expensive. It is specifically attractive for sensors that are wideband, wide-angle, and/or have long coherent apertures with substantial off-track motion.

More Complex Operation

The basic design of a synthetic aperture radar system can be enhanced to collect more information. Most of these methods use the same basic principle of combining many pulses to form a synthetic aperture, but may involve additional antennas or significant additional processing.

Multistatic Operation

SAR requires that echo captures be taken at multiple antenna positions. The more captures taken (at different antenna locations) the more reliable the target characterization.

Multiple captures can be obtained by moving a single antenna to different locations, by placing multiple stationary antennas at different locations, or combinations thereof.

The advantage of a single moving antenna is that it can be easily placed in any number of positions to provide any number of monostatic waveforms. For example, an antenna mounted on an airplane takes many captures per second as the plane travels.

The principal advantages of multiple static antennas are that a moving target can be characterized (assuming the capture electronics are fast enough), that no vehicle or motion machinery is necessary, and that antenna positions need not be derived from other, sometimes unreliable, information. (One problem with SAR aboard an airplane is knowing precise antenna positions as the plane travels).

For multiple static antennas, all combinations of monostatic and multistatic radar waveform captures are possible. Note, however, that it is not advantageous to capture a waveform for each of both transmission directions for a given pair of antennas, because those waveforms will be identical. When multiple static antennas are used, the total number of unique echo waveforms that can be captured is:

$$\frac{N^2 + N}{2}$$

where N is the number of unique antenna positions.

Modes

Stripmap Mode Airborne SAR

The antenna stays in a fixed position, and may be orthogonal to the flight path or squinted slightly forward or backward .

When the antenna aperture travels along the flight path, a signal is transmitted at a rate equal to the pulse repetition frequency (PRF). The lower boundary of the PRF is determined by the Doppler bandwidth of the radar. The backscatter of each of these signals is commutatively added on a pixel-by-pixel basis to attain the fine azimuth resolution desired in radar imagery.

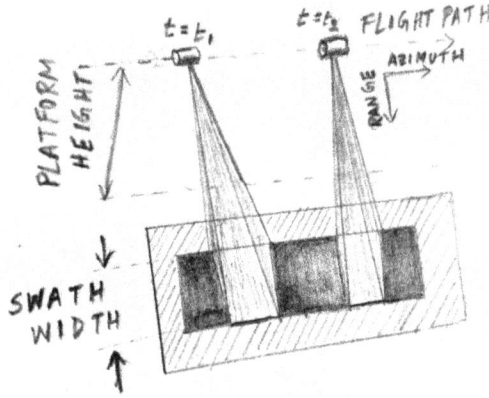

Illustration of the SAR stripmap operation mode.

Spotlight Mode SAR

The spotlight synthetic aperture is given by-

$$Lsa = r_0.\Delta\theta_a$$

where $\Delta\theta_a$ is the angle formed between the beginning and end of the imaging, as shown in the diagram of spotlight imaging and r_0 is the range distance.

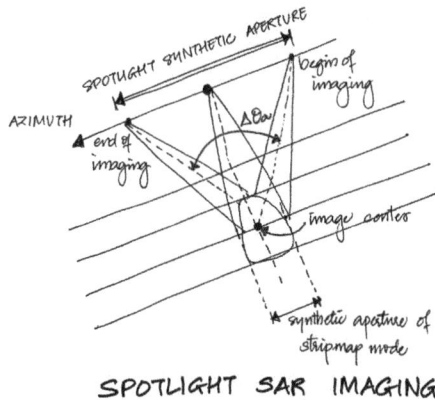

Depiction of the Spotlight Image Mode

The spotlight mode gives better resolution for a smaller ground patch. In this mode, the illuminating radar beam is steered continually as the aircraft moves, so that it illuminates the same patch over a longer period of time. This mode is not a very continuous imaging mode; however, has high azimuth resolution.

Scan Mode SAR

While operating as a scan mode SAR, the antenna beam sweeps periodically and thus cover much larger area than spotlight and stripmap modes. However, the azimuth resolution become much lower than stripmap mode due to decreased azimuth bandwidth. Clearly there is a balance achieved between azimuth resolution and scan area of SAR. Here, the synthetic aperture is shared between the sub swaths, and it is not in direct contact within one subswath. Mosaic Operation is required in Azimuth and range directions to join the azimuth bursts and the range sub-swaths.

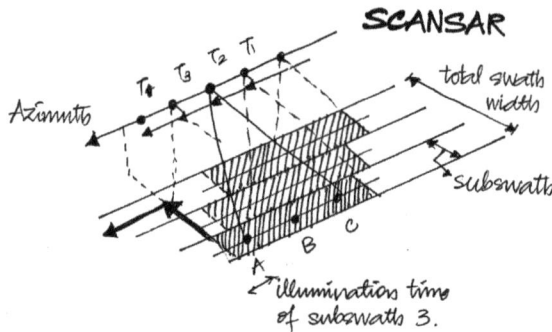

Depiction of ScanSAR Imaging Mode

Properties

- ScanSAR makes the swath beam huge.

- The azimuth signal has many bursts.

- The Azimuth resolution is limited due to the burst duration.

- Each target contains varied frequency which completely depends where the Azimuth is present.

Polarimetry

Radar waves have a polarization. Different materials reflect radar waves with different intensities, but anisotropic materials such as grass often reflect different polarizations with different intensities. Some materials will also convert one polarization into another. By emitting a mixture of polarizations and using receiving antennas with a specific polarization, several images can be collected from the same series of pulses. Frequently three such RX-TX polarizations (HH-pol, VV-pol, VH-pol) are used as the three color channels in a synthesized image. This is what has been done in the picture at right. Interpretation of the resulting colors requires significant testing of known materials.

SAR image of Death Valley colored using polarimetry

New developments in polarimetry include using the changes in the random polarization returns of some surfaces (such as grass or sand) and between two images of the same location at different times to determine where changes not visible to optical systems occurred. Examples include subterranean tunneling or paths of vehicles driving through the area being imaged. Enhanced SAR sea oil slick observation has been developed by appropriate physical modelling and use of fully polarimetric and dual-polarimetric measurements.

SAR polarimetry is a technique used for deriving qualitative and quantitative physical information for land, snow and ice, ocean and urban applications based on the measurement and exploration of the polarimetric properties of man-made and natural scatterers. *Terrain* and *land use* classification is one of the most important applications of polarimetric synthetic aperture radar (POLSAR).

SAR polarimetry uses a scattering matrix (S) to identify the scattering behavior of objects after an interaction with electromagnetic wave. The matrix is represented by a combination of horizontal and vertical polarization states of transmitted and received signals.

$$S = \begin{bmatrix} S_{HH} & S_{HV} \\ S_{VH} & S_{VV} \end{bmatrix}$$

where, HH is for horizontal transmit and horizontal receive, VV is for vertical transmit and vertical receive, HV is for horizontal transmit and vertical receive, and VH – for vertical transmit and horizontal receive.

The first two of these polarization combinations are referred to as like-polarized, because the transmit and receive polarizations are the same. The last two combinations are referred to as cross-polarized because the transmit and receive polarizations are orthogonal to one another.

The three-component scattering power model by Freeman and Durden is successfully used for decomposition of POLSAR image, applying the reflection symmetry condition using covariance matrix. The method is based on simple physical scattering mechanisms (surface scattering, double-bounce scattering, and volume scattering). The advantage of this scattering model is that it is simple and easy to implement for image processing. There are 2 major approaches for a 3X3 polarimetric matrix decomposition. One is the lexicographic covariance matrix approach based on physically measurable parameters, and the other is the Pauli decomposition which is a coherent decomposition matrix. It represents all the polarimetric information in a single SAR image. The polarimetric information of [S] could be represented by the combination of the intensities $|S_{HH}|^2, |S_{VV}|^2, 2|S_{HV}|^2$ in a single RGB image where all the previous intensities will be coded as a color channel.

For PolSAR image analysis, there can be cases where reflection symmetry condition does not hold. In those cases a *four-component scattering model* can be used to decompose polarimetric synthetic aperture radar (SAR) images. This approach deals with the non- reflection symmetric scattering case. It includes and extends the three-component decomposition method introduced by Freeman and Durden to a fourth component by adding the helix scattering power. This helix power term generally appears in complex urban area but disappears for a natural distributed scatterer.

There is also an improved method using the four-component decomposition algorithm, which was introduced for the general POLSAR data image analyses. The SAR data is first filtered which is known as speckle reduction, then each pixel is decomposed by four-component model to determine the surface scattering power (P_s), double-bounce scattering power (P_d), volume scattering power (P_v), and helix scattering power (P_c). The pixels are then divided into 5 classes (surface,double-bounce,volume,helix,and mixed pixels) classified with respect to maximum powers. A mixed category is added for the pixels having two or three equal dominant scattering powers after computation. The process continues as the pixels in all these categories are divided in 20 small clutter approximately of same number of pixels and merged as desirable, this is called cluster merging. They are iteratively classified and then automatically color is delivered to each class. The summarization of this algorithm leads to an understanding that, brown colors denotes the surface scattering classes, red colors for double-bounce scattering classes, green colors for volume scattering classes, and blue colors for helix scattering classes.

Color representation of different polarizations.

Although this method is aimed for non-reflection case, it automatically includes the reflection symmetry condition, therefore in can be used as a general case. It also preserves the scattering characteristics by taking the mixed scattering category into account therefore proving to be a better algorithm.

Interferometry

Rather than discarding the phase data, information can be extracted from it. If two observations of the same terrain from very similar positions are available, aperture synthesis can be performed to provide the resolution performance which would be given by a radar system with dimensions equal to the separation of the two measurements. This technique is called interferometric SAR or InSAR.

If the two samples are obtained simultaneously (perhaps by placing two antennas on the same aircraft, some distance apart), then any phase difference will contain information about the angle from which the radar echo returned. Combining this with the distance information, one can determine the position in three dimensions of the image pixel. In other words, one can extract terrain altitude as well as radar reflectivity, producing a digital elevation model (DEM) with a single airplane pass. One aircraft application at the Canada Centre for Remote Sensing produced digital elevation maps with a resolution of 5 m and altitude errors also about 5 m. Interferometry was used to map many regions of the Earth's surface with unprecedented accuracy using data from the Shuttle Radar Topography Mission.

If the two samples are separated in time, perhaps from two flights over the same terrain, then there are two possible sources of phase shift. The first is terrain altitude, as discussed above. The second is terrain motion: if the terrain has shifted between observations, it will return a different phase. The amount of shift required to cause a significant phase difference is on the order of the wavelength used. This means that if the terrain shifts by centimeters, it can be seen in the resulting image (a digital elevation map must be available to separate the two kinds of phase difference; a third pass may be necessary to produce one).

This second method offers a powerful tool in geology and geography. Glacier flow can be mapped with two passes. Maps showing the land deformation after a minor earthquake or after a volcanic eruption (showing the shrinkage of the whole volcano by several centimeters) have been published (where?).

Differential Interferometry

Differential interferometry (D-InSAR) requires taking at least two images with addition of a DEM. The DEM can be either produced by GPS measurements or could be generated by interferometry as long as the time between acquisition of the image pairs is short, which guarantees minimal distortion of the image of the target surface. In principle, 3 images of the ground area with similar image acquisition geometry is often adequate for D-InSar. The principle for detecting ground movement is quite simple. One interferogram is created from the first two images; this is also called the reference interferogram or topographical interferogram. A second interferogram is created that captures topography + distortion. Subtracting the latter from the reference interferogram can reveal differential fringes, indicating movement. The described 3 image D-InSAR generation technique is called 3-pass or double-difference method.

Differential fringes which remain as fringes in the differential interferogram are a result of SAR range changes of any displaced point on the ground from one interferogram to the next. In the differential interferogram, each fringe is directly proportional to the SAR wavelength, which is about 5.6 cm for ERS and RADARSAT single phase cycle. Surface displacement away from the satellite look direction causes an increase in path (translating to phase) difference. Since the signal travels from the SAR antenna to the target and back again, the measured displacement is twice the unit of wavelength. This means in differential interferometry one fringe cycle $-\pi$ to $+\pi$ or one wavelength corresponds to a displacement relative to SAR antenna of only half wavelength (2.8 cm). There are various publications on measuring subsidence movement, slope stability analysis, landslide, glacier movement, etc. tooling D-InSAR. Further advancement to this technique whereby differential interferometry from satellite SAR ascending pass and descending pass can be used to estimate 3-D ground movement. Research in this area has shown accurate measurements of 3-D ground movement with accuracies comparable to GPS based measurements can be achieved.

Tomo-SAR

SAR Tomography is a subfield of a concept named as multi-baseline interferometry. It has been developed to give a 3D exposure to the imaging, which uses the beam formation concept. It can be used when the use demands a focused phase concern between the magnitude and the phase components of the SAR data, during information retrieval. One of the major advantages of Tomo-SAR is that it can separate out the parameters which get scattered, irrespective of how different their motions are.

On using Tomo-SAR with differential interferometry, a new combination named "differential tomography" (Diff-Tomo) is developed.

Application of Tomo-SAR

Tomo-SAR has an application based on radar imaging, which is the depiction of Ice Volume and Forest Temporal Coherence (Temporal coherence describes the correlation between waves observed at different moments in time).

Ultra-wideband SAR

Conventional radar systems emit bursts of radio energy with a fairly narrow range of frequencies. A narrow-band channel, by definition, does not allow rapid changes in modulation. Since it is the change in a received signal that reveals the time of arrival of the signal (obviously an unchanging signal would reveal nothing about "when" it reflected from the target), a signal with only a slow change in modulation cannot reveal the distance to the target as well as can a signal with a quick change in modulation.

Ultra-wideband (UWB) refers to any radio transmission that uses a very large bandwidth – which is the same as saying it uses very rapid changes in modulation. Although there is no set bandwidth value that qualifies a signal as "UWB", systems using bandwidths greater than a sizable portion of the center frequency (typically about ten percent, or so) are most often called "UWB" systems. A typical UWB system might use a bandwidth of one-third to one-half of its center frequency. For example, some systems use a bandwidth of about 1 GHz centered around 3 GHz.

There are as many ways to increase the bandwidth of a signal as there are forms of modulation – it is simply a matter of increasing the rate of that modulation. However, the two most common methods used in UWB radar, including SAR, are very short pulses and high-bandwidth chirping. A general description of chirping appears elsewhere in this article. The bandwidth of a chirped system can be as narrow or as wide as the designers desire. Pulse-based UWB systems, being the more common method associated with the term "UWB radar", are described here.

A pulse-based radar system transmits very short pulses of electromagnetic energy, typically only a few waves or less. A very short pulse is, of course, a very rapidly changing signal, and thus occupies a very wide bandwidth. This allows far more accurate measurement of distance, and thus resolution.

The main disadvantage of pulse-based UWB SAR is that the transmitting and receiving front-end electronics are difficult to design for high-power applications. Specifically, the transmit duty cycle is so exceptionally low and pulse time so exceptionally short, that the electronics must be capable of extremely high instantaneous power to rival the average power of conventional radars. (Although it is true that UWB provides a notable gain in channel capacity over a narrow band signal because of the relationship of bandwidth in the Shannon–Hartley theorem and because the low receive duty cycle receives less noise, increasing the signal-to-noise ratio, there is still a notable disparity in link budget because conventional radar might be several orders of magnitude more powerful than a typical pulse-based radar.) So pulse-based UWB SAR is typically used in applications requiring average power levels in the microwatt or milliwatt range, and thus is used for scanning smaller, nearer target areas (several tens of meters), or in cases where lengthy integration (over a span of minutes) of the received signal is possible. Note, however, that this limitation is solved in chirped UWB radar systems.

The principal advantages of UWB radar are better resolution (a few millimeters using commercial off-the-shelf electronics) and more spectral information of target reflectivity.

Doppler-beam Sharpening

Doppler Beam Sharpening commonly refers to the method of processing unfocused real-beam phase history to achieve better resolution than could be achieved by processing the real beam without it. Because the real aperture of the radar antenna is so small (compared to the wavelength in use), the radar energy spreads over a wide area (usually many degrees wide in a direction orthogonal (at right angles) to the direction of the platform (aircraft)). Doppler-beam sharpening takes advantage of the motion of the platform in that targets ahead of the platform return a Doppler upshifted signal (slightly higher in frequency) and targets behind the platform return a Doppler downshifted signal (slightly lower in frequency).

The amount of shift varies with the angle forward or backward from the ortho-normal direction. By knowing the speed of the platform, target signal return is placed in a specific angle "bin" that changes over time. Signals are integrated over time and thus the radar "beam" is synthetically reduced to a much smaller aperture – or more accurately (and based on the ability to distinguish smaller Doppler shifts) the system can have hundreds of very "tight" beams concurrently. This technique dramatically improves angular resolution; however, it is far more difficult to take advantage of this technique for range resolution.

Chirped (Pulse-compressed) Radars

A common technique for many radar systems (usually also found in SAR systems) is to "chirp" the signal. In a "chirped" radar, the pulse is allowed to be much longer. A longer pulse allows more energy to be emitted, and hence received, but usually hinders range resolution. But in a chirped radar, this longer pulse also has a frequency shift during the pulse (hence the chirp or frequency shift). When the "chirped" signal is returned, it must be correlated with the sent pulse. Classically, in analog systems, it is passed to a dispersive delay line (often a SAW device) that has the property of varying velocity of propagation based on frequency. This technique "compresses" the pulse in time – thus having the effect of a much shorter pulse (improved range resolution) while having the benefit of longer pulse length (much more signal returned). Newer systems use digital pulse correlation to find the pulse return in the signal.

Typical Operation

NASA's AirSAR instrument is attached to the side of a DC-8

In a typical SAR application, a single radar antenna is attached to an aircraft or space-craft so as to radiate a beam whose wave-propagation direction has a substantial com-ponent perpendicular to the flight-path direction. The beam is allowed to be broad in the vertical direction so it will illuminate the terrain from nearly beneath the aircraft out toward the horizon.

Resolution in the range dimension of the image is accomplished by creating pulses which define very short time intervals, either by emitting short pulses consisting of a carrier frequency and the necessary sidebands, all within a certain bandwidth, or by using longer "chirp pulses" in which frequency varies (often linearly) with time within that bandwidth. The differing times at which echoes return allow points at different distances to be distinguished.

The total signal is that from a beamwidth-sized patch of the ground. To produce a beam that is narrow in the cross-range direction, diffraction effects require that the antenna be wide in that dimension. Therefore, the distinguishing, from each other, of co-range points simply by strengths of returns that persist for as long as they are within the beam width is difficult with aircraft-carryable antennas, because their beams can have linear widths only about two orders of magnitude (hundreds of times) smaller than the range. (Spacecraft-carryable ones can do 10 or more times better.) However, if both the amplitude and the phase of returns are recorded, then the portion of that multi-target return that was scattered radially from any smaller scene element can be extracted by phase-vector correlation of the total return with the form of the return expected from each such element. Careful design and operation can accomplish resolution of items smaller than a millionth of the range, for example, 30 cm at 300 km, or about one foot at nearly 200 miles (320 km).

The process can be thought of as combining the series of spatially distributed observations as if all had been made simultaneously with an antenna as long as the beamwidth and focused on that particular point. The "synthetic aperture" simulated at maximum system range by this process not only is longer than the real antenna, but, in practical applications, it is much longer than the radar aircraft, and tremendously longer than the radar spacecraft.

Image resolution of SAR in its range coordinate (expressed in image pixels per distance unit) is mainly proportional to the radio bandwidth of whatever type of pulse is used. In the cross-range coordinate, the similar resolution is mainly proportional to the bandwidth of the Doppler shift of the signal returns within the beamwidth. Since Doppler frequency depends on the angle of the scattering point's direction from the broadside direction, the Doppler bandwidth available within the beamwidth is the same at all ranges. Hence the theoretical spatial resolution limits in both image dimensions remain constant with variation of range. However, in practice, both the errors that accumulate with data-collection time and the particular techniques used in post-processing further limit cross-range resolution at long ranges.

The conversion of return delay time to geometric range can be very accurate because of the natural constancy of the speed and direction of propagation of electromagnetic waves. However, for an aircraft flying through the never-uniform and never-quiescent atmosphere, the relating of pulse transmission and reception times to successive geometric positions of the antenna must be accompanied by constant adjusting of the return phases to account for sensed irregularities in the flight path. SAR's in spacecraft avoid that atmosphere problem, but still must make corrections for known antenna movements due to rotations of the spacecraft, even those that are reactions to movements of onboard machinery. Locating a SAR in a manned space vehicle may require that the humans carefully remain motionless relative to the vehicle during data collection periods.

Although some references to SARs have characterized them as "radar telescopes", their actual optical analogy is the microscope, the detail in their images being smaller than the length of the synthetic aperture. In radar-engineering terms, while the target area is in the "far field" of the illuminating antenna, it is in the "near field" of the simulated one.

Returns from scatterers within the range extent of any image are spread over a matching time interval. The inter-pulse period must be long enough to allow farthest-range returns from any pulse to finish arriving before the nearest-range ones from the next pulse begin to appear, so that those do not overlap each other in time. On the other hand, the interpulse rate must be fast enough to provide sufficient samples for the desired across-range (or across-beam) resolution. When the radar is to be carried by a high-speed vehicle and is to image a large area at fine resolution, those conditions may clash, leading to what has been called SAR's ambiguity problem. The same considerations apply to "conventional" radars also, but this problem occurs significantly only when resolution is so fine as to be available only through SAR processes. Since the basis of the problem is the information-carrying capacity of the single signal-input channel provided by one antenna, the only solution is to use additional channels fed by additional antennas. The system then becomes a hybrid of a SAR and a phased array, sometimes being called a Vernier array.

Combining the series of observations requires significant computational resources, usually using Fourier transform techniques. The high digital computing speed now available allows such processing to be done in near-real time on board a SAR aircraft. (There is necessarily a minimum time delay until all parts of the signal have been received.) The result is a map of radar reflectivity, including both amplitude and phase. The amplitude information, when shown in a map-like display, gives information about ground cover in much the same way that a black-and-white photo does. Variations in processing may also be done in either vehicle-borne stations or ground stations for various purposes, so as to accentuate certain image features for detailed target-area analysis.

Although the phase information in an image is generally not made available to a human observer of an image display device, it can be preserved numerically, and sometimes allows certain additional features of targets to be recognized. Unfortunately, the phase differences between adjacent image picture elements ("pixels") also produce random interference effects called "coherence speckle", which is a sort of graininess with dimensions on the order of the resolution, causing the concept of resolution to take on a subtly different meaning. This effect is the same as is apparent both visually and photographically in laser-illuminated optical scenes. The scale of that random speckle structure is governed by the size of the synthetic aperture in wavelengths, and cannot be finer than the system's resolution. Speckle structure can be subdued at the expense of resolution.

Before rapid digital computers were available, the data processing was done using an optical holography technique. The analog radar data were recorded as a holographic interference pattern on photographic film at a scale permitting the film to preserve the signal bandwidths (for example, 1:1,000,000 for a radar using a 0.6-meter wavelength). Then light using, for example, 0.6-micrometer waves (as from a helium–neon laser) passing through the hologram could project a terrain image at a scale recordable on another film at reasonable processor focal distances of around a meter. This worked because both SAR and phased arrays are fundamentally similar to optical holography, but using microwaves instead of light waves. The "optical data-processors" developed for this radar purpose were the first effective analog optical computer systems, and were, in fact, devised before the holographic technique was fully adapted to optical imaging. Because of the different sources of range and across-range signal structures in the radar signals, optical data-processors for SAR included not only both spherical and cylindrical lenses, but sometimes conical ones.

Image Appearance

The following considerations apply also to real-aperture terrain-imaging radars, but are more consequential when resolution in range is matched to a cross-beam resolution that is available only from a SAR.

The two dimensions of a radar image are range and cross-range. Radar images of limited patches of terrain can resemble oblique photographs, but not ones taken from the location of the radar. This is because the range coordinate in a radar image is perpendicular to the vertical-angle coordinate of an oblique photo. The apparent entrance-pupil position (or camera center) for viewing such an image is therefore not as if at the radar, but as if at a point from which the viewer's line of sight is perpendicular to the slant-range direction connecting radar and target, with slant-range increasing from top to bottom of the image.

Because slant ranges to level terrain vary in vertical angle, each elevation of such terrain appears as a curved surface, specifically a hyperbolic cosine one. Verticals at various ranges are perpendiculars to those curves. The viewer's apparent looking directions are parallel to the curve's "hypcos" axis. Items directly beneath the radar appear as if optically viewed horizontally (i.e., from the side) and those at far ranges as if optically viewed from directly above. These curvatures are not evident unless large extents of near-range terrain, including steep slant ranges, are being viewed.

When viewed as specified above, fine-resolution radar images of small areas can appear most nearly like familiar optical ones, for two reasons. The first reason is easily understood by imagining a flagpole in the scene. The slant-range to its upper end is less than that to its base. Therefore, the pole can appear correctly top-end up only when viewed in the above orientation. Secondly, the radar illumination then being downward, shadows are seen in their most-familiar "overhead-lighting" direction.

Note that the image of the pole's top will overlay that of some terrain point which is on the same slant range arc but at a shorter horizontal range ("ground-range"). Images of scene surfaces which faced both the illumination and the apparent eyepoint will have geometries that resemble those of an optical scene viewed from that eyepoint. However, slopes facing the radar will be foreshortened and ones facing away from it will be lengthened from their horizontal (map) dimensions. The former will therefore be brightened and the latter dimmed.

Returns from slopes steeper than perpendicular to slant range will be overlaid on those of lower-elevation terrain at a nearer ground-range, both being visible but intermingled. This is especially the case for vertical surfaces like the walls of buildings. Another viewing inconvenience that arises when a surface is steeper than perpendicular to the slant range is that it is then illuminated on one face but "viewed" from the reverse face. Then one "sees", for example, the radar-facing wall of a building as if from the inside, while the building's interior and the rear wall (that nearest to, hence expected to be optically visible to, the viewer) have vanished, since they lack illumination, being in the shadow of the front wall and the roof. Some return from the roof may overlay that from the front wall, and both of those may overlay return from terrain in front of the building. The visible building shadow will include those of all illuminated items. Long shadows may exhibit blurred edges due to the illuminating antenna's movement during the "time exposure" needed to create the image.

Surfaces that we usually consider rough will, if that roughness consists of relief less than the radar wavelength, behave as smooth mirrors, showing, beyond such a surface, additional images of items in front of it. Those mirror images will appear within the shadow of the mirroring surface, sometimes filling the entire shadow, thus preventing recognition of the shadow.

An important fact that applies to SARs but not to real-aperture radars is that the direction of overlay of any scene point is not directly toward the radar, but toward that point of the SAR's current path direction that is nearest to the target point. If the SAR is "squinting" forward or aft away from the exactly broadside direction, then the illumination direction, and hence the shadow direction, will not be opposite to the overlay direction, but slanted to right or left from it. An image will appear with the correct projection geometry when viewed so that the overlay direction is vertical, the SAR's flight-path is above the image, and range increases somewhat downward.

Objects in motion within a SAR scene alter the Doppler frequencies of the returns. Such objects therefore appear in the image at locations offset in the across-range direction by amounts proportional to the range-direction component of their velocity. Road vehicles may be depicted off the roadway and therefore not recognized as road traffic items. Trains appearing away from their tracks are more easily properly recognized by their length parallel to known trackage as well as by the absence of an equal length of railbed signature and of some adjacent terrain, both having been shadowed by the train. While images of moving vessels can be offset from the line of the earlier parts of their wakes, the more recent parts of the wake, which still partake of some of the vessel's motion, appear as curves connecting the vessel image to the relatively quiescent far-aft wake. In such identifiable cases, speed and direction of the moving items can be determined from the amounts of their offsets. The along-track component of a target's motion causes some defocus. Random motions such as that of wind-driven tree foliage, vehicles driven over rough terrain, or humans or other animals walking or running generally render those items not focusable, resulting in blurring or even effective invisibility.

These considerations, along with the speckle structure due to coherence, take some getting used to in order to correctly interpret SAR images. To assist in that, large collections of significant target signatures have been accumulated by performing many test flights over known terrains and cultural objects.

History

Carl A. Wiley, a mathematician at Goodyear Aircraft Company in Litchfield Park, Arizona, invented synthetic aperture radar in June 1951 while working on a correlation guidance system for the Atlas ICBM program. In early 1952, Wiley, together with Fred Heisley and Bill Welty, constructed a concept validation system known as DOUSER ("Doppler Unbeamed Search Radar"). During the 1950s and 1960s, Goodyear Aircraft (later Goodyear Aerospace) introduced numerous advancements in SAR technology, many with the help from Don Beckerleg.

Independently of Wiley's work, experimental trials in early 1952 by Sherwin and others at the University of Illinois' Control Systems Laboratory showed results that they pointed out "could provide the basis for radar systems with greatly improved angular resolution" and might even lead to systems capable of focusing at all ranges simultaneously.

In both of those programs, processing of the radar returns was done by electrical-circuit filtering methods. In essence, signal strength in isolated discrete bands of Doppler frequency defined image intensities that were displayed at matching angular positions within proper range locations. When only the central (zero-Doppler band) portion of the return signals was used, the effect was as if only that central part of the beam existed. That led to the term Doppler Beam Sharpening. Displaying returns from several adjacent non-zero Doppler frequency bands accomplished further "beam-subdividing" (sometimes called "unfocused radar", though it could have been considered "semi-focused"). Wiley's patent, applied for in 1954, still proposed similar processing. The bulkiness of the circuitry then available limited the extent to which those schemes might further improve resolution.

The principle was included in a memorandum authored by Walter Hausz of General Electric that was part of the then-secret report of a 1952 Dept. of Defense summer study conference called TEOTA ("The Eyes of the Army"), which sought to identify new techniques useful for military reconnaissance and technical gathering of intelligence. A follow-on summer program in 1953 at the University of Michigan, called Project Wolverine, identified several of the TEOTA subjects, including Doppler-assisted sub-beam-width resolution, as research efforts to be sponsored by the Department of Defense (DoD) at various academic and industrial research laboratories. In that same year, the Illinois group produced a "strip-map" image exhibiting a considerable amount of sub-beamwidth resolution.

A more advanced focused-radar project was among several remote sensing schemes assigned in 1953 to Project Michigan, a tri-service-sponsored (Army, Navy, Air Force) program at the University of Michigan's Willow Run Research Center (WRRC), that program being administered by the Army Signal Corps. Initially called the side-looking radar project, it was carried out by a group first known as the Radar Laboratory and later as the Radar and Optics Laboratory. It proposed to take into account, not just the short-term existence of several particular Doppler shifts, but the entire history of the steadily varying shifts from each target as the latter crossed the beam. An early analysis by Dr. Louis J. Cutrona, Weston E. Vivian, and Emmett N. Leith of that group showed that such a fully focused system should yield, at all ranges, a resolution equal to the width (or, by some criteria, the half-width) of the real antenna carried on the radar aircraft and continually pointed broadside to the aircraft's path.

The required data processing amounted to calculating cross-correlations of the received signals with samples of the forms of signals to be expected from unit-amplitude sources at the various ranges. At that time, even large digital computers had capabilities somewhat near the levels of today's four-function handheld calculators, hence were nowhere near able to do such a huge amount of computation. Instead, the device for doing the correlation computations was to be an optical correlator.

It was proposed that signals received by the traveling antenna and coherently detected be displayed as a single range-trace line across the diameter of the face of a cathode-ray tube, the line's successive forms being recorded as images projected onto a film traveling perpendicular to the length of that line. The information on the developed film was to be subsequently processed in the laboratory on equipment still to be devised as a principal task of the project. In the initial processor proposal, an arrangement of lenses was expected to multiply the recorded signals point-by-point with the known signal forms by passing light successively through both the signal film and another film containing the known signal pattern. The subsequent summation, or integration, step of the correlation was to be done by converging appropriate sets of multiplication products by the focusing action of one or more spherical and cylindrical lenses. The processor was to be, in effect, an optical analog computer performing large-scale scalar arithmetic calculations in many channels (with many light "rays") at once. Ultimately, two such devices would be needed, their outputs to be combined as quadrature components of the complete solution.

Fortunately (as it turned out), a desire to keep the equipment small had led to recording the reference pattern on 35 mm film. Trials promptly showed that the patterns on the film were so fine as to show pronounced diffraction effects that prevented sharp final focusing.

That led Leith, a physicist who was devising the correlator, to recognize that those effects in themselves could, by natural processes, perform a significant part of the needed processing, since along-track strips of the recording operated like diametrical slices of a series of circular optical zone plates. Any such plate performs somewhat like a lens, each plate having a specific focal length for any given wavelength. The recording that had been considered as scalar became recognized as pairs of opposite-sign vector ones of many spatial frequencies plus a zero-frequency "bias" quantity. The needed correlation summation changed from a pair of scalar ones to a single vector one.

Each zone plate strip has two equal but oppositely signed focal lengths, one real, where a beam through it converges to a focus, and one virtual, where another beam appears to have diverged from, beyond the other face of the zone plate. The zero-frequency (DC bias) component has no focal point, but overlays both the converging and diverging beams. The key to obtaining, from the converging wave component, focused images that are not overlaid with unwanted haze from the other two is to block the latter, allowing only the wanted beam to pass through a properly positioned frequency-band selecting aperture.

Each radar range yields a zone plate strip with a focal length proportional to that range. This fact became a principal complication in the design of optical processors. Consequently, technical journals of the time contain a large volume of material devoted to ways for coping with the variation of focus with range.

For that major change in approach, the light used had to be both monochromatic and coherent, properties that were already a requirement on the radar radiation. Lasers also then being in the future, the best then-available approximation to a coherent light source was the output of a mercury vapor lamp, passed through a color filter that was matched to the lamp spectrum's green band, and then concentrated as well as possible onto a very small beam-limiting aperture. While the resulting amount of light was so weak that very long exposure times had to be used, a workable optical correlator was assembled in time to be used when appropriate data became available.

Although creating that radar was a more straightforward task based on already-known techniques, that work did demand the achievement of signal linearity and frequency stability that were at the extreme state of the art. An adequate instrument was designed and built by the Radar Laboratory and was installed in a C-46 (Curtiss Commando) aircraft. Because the aircraft was bailed to WRRC by the U. S. Army and was flown and maintained by WRRC's own pilots and ground personnel, it was available for many flights at times matching the Radar Laboratory's needs, a feature important for allowing frequent re-testing and "debugging" of the continually developing complex equipment. By contrast, the Illinois group had used a C-46 belonging to the Air Force and flown by AF pilots only by pre-arrangement, resulting, in the eyes of those researchers, in limitation to a less-than-desirable frequency of flight tests of their equipment, hence a low bandwidth of feedback from tests. (Later work with newer Convair aircraft continued the Michigan group's local control of flight schedules.)

Michigan's chosen 5-foot (1.5 m)-wide World War II-surplus antenna was theoretically capable of 5-foot (1.5 m) resolution, but data from only 10% of the beamwidth was used at first, the goal at that time being to demonstrate 50-foot (15 m) resolution. It was understood that finer resolution would require the added development of means for sensing departures of the aircraft from an ideal heading and flight path, and for using that information for making needed corrections to the antenna pointing and to the received signals before processing. After numerous trials in which even small atmospheric turbulence kept the aircraft from flying straight and level enough for good 50-foot (15 m) data, one pre-dawn flight in August 1957 yielded a map-like image of the Willow Run Airport area which did demonstrate 50-foot (15 m) resolution in some parts of the image, whereas the illuminated beam width there was 900 feet (270 m). Although the program had been considered for termination by DoD due to what had seemed to be a lack of results, that first success ensured further funding to continue development leading to solutions to those recognized needs.

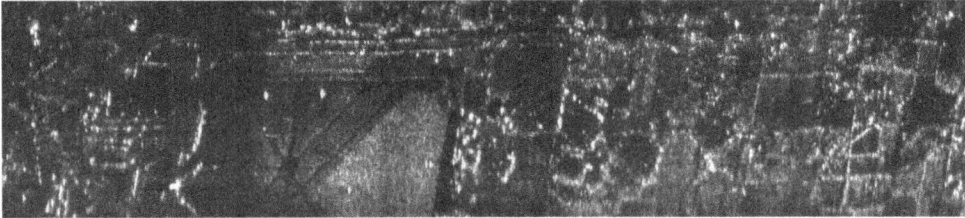

First successful focussed airborne synthetic aperture radar image, Willow Run Airport and vicinity, August 1957. Image courtesy University of Michigan.

The SAR principle was first acknowledged publicly via an April 1960 press release about the U. S. Army experimental AN/UPD-1 system, which consisted of an airborne element made by Texas Instruments and installed in a Beech L-23D aircraft and a mobile ground data-processing station made by WRRC and installed in a military van. At the time, the nature of the data processor was not revealed. A technical article in the journal of the IRE (Institute of Radio Engineers) Professional Group on Military Electronics in February 1961 described the SAR principle and both the C-46 and AN/UPD-1 versions, but did not tell how the data were processed, nor that the UPD-1's maximum resolution capability was about 50 feet (15 m). However, the June 1960 issue of the IRE Professional Group on Information Theory had contained a long article on "Optical Data Processing and Filtering Systems" by members of the Michigan group. Although it did not refer to the use of those techniques for radar, readers of both journals could quite easily understand the existence of a connection between articles sharing some authors.

An operational system to be carried in a reconnaissance version of the F-4 "Phantom" aircraft was quickly devised and was used briefly in Vietnam, where it failed to favorably impress its users, due to the combination of its low resolution (similar to the UPD-1's), the speckly nature of its coherent-wave images (similar to the speckliness of laser images), and the poorly understood dissimilarity of its range/cross-range images from the angle/angle optical ones familiar to military photo interpreters. In subsequent work the technique's latent capability was eventually achieved. That work, depending on advanced radar circuit designs and precision sensing of departures from ideal straight flight, along with more sophisticated optical processors using laser light sources and specially designed very large lenses made from remarkably clear glass, allowed the Michigan group to advance system resolution, at about 5-year intervals, first to 15 feet (4.6 m), then 5 feet (1.5 m), and, by the mid-1970s, to 1 foot (the latter only over very short range intervals while processing was still being done optically). The latter levels and the associated very wide dynamic range proved suitable for identifying many objects of military concern as well as soil, water, vegetation, and ice features being studied by a variety of environmental researchers having security clearances allowing them access to what was then classified imagery. Similarly improved operational systems soon followed each of those finer-resolution steps.

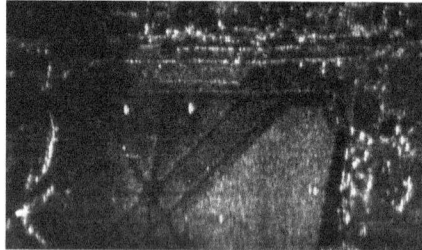

Portion of the first successful optically processed SAR image, showing Willow Run Airport, Aug 1957. Resolution varied from 50 feet to 100 feet within a beam 900 feet wide. The two large spot returns were from large corner reflectors used as test targets, appearing over-large here because their signals exceeded the dynamic range of the recording film. Widths of runways were about 150 feet. Range and cross-range scales not matched nor at right angles.

Portion of a later 15-foot SAR image, showing a part of Willow Run Airport that appeared in the upper image. The very bright line represents the airport's chain-link boundary fence. Above it is a divided highway with two paved lanes in each direction, and above that is the undivided two lanes of a paved frontage road. Note the speckle structure, which is an essential feature of coherent-wave images.

Comparison of earliest SAR image with a later improved-resolution one. Additionally, the data-processing light source had been changed from a mercury lamp to a laser. Image data courtesy of University of Michigan and Natural Resources Canada.

Even the 5-foot (1.5 m) resolution stage had over-taxed the ability of cathode-ray tubes (limited to about 2000 distinguishable items across the screen diameter) to deliver fine enough details to signal films while still covering wide range swaths, and taxed the optical processing systems in similar ways. However, at about the same time, digital computers finally became capable of doing the processing without similar limitation, and the consequent presentation of the images on cathode ray tube monitors instead of film allowed for better control over tonal reproduction and for more convenient image mensuration.

Achievement of the finest resolutions at long ranges was aided by adding the capability to swing a larger airborne antenna so as to more strongly illuminate a limited target area continually while collecting data over several degrees of aspect, removing the previous limitation of resolution to the antenna width. This was referred to as the spotlight mode, which no longer produced continuous-swath images but, instead, images of isolated patches of terrain.

It was understood very early in SAR development that the extremely smooth orbital path of an out-of-the-atmosphere platform made it ideally suited to SAR operation. Early experience with artificial earth satellites had also demonstrated that the Doppler frequency shifts of signals traveling through the ionosphere and atmosphere were stable enough to permit very fine resolution to be achievable even at ranges of hundreds of kilometers. While further experimental verification of those facts by a project now referred to as the Quill satellite (declassified in 2012) occurred within the second decade after the initial work began, several of the capabilities for creating useful classified systems did not exist for another two decades.

That seemingly slow rate of advances was often paced by the progress of other inventions, such as the laser, the digital computer, circuit miniaturization, and compact data storage. Once the laser appeared, optical data processing became a fast process because it provided many parallel analog channels, but devising optical chains suited to matching signal focal lengths to ranges proceeded by many stages and turned out to call for some novel optical components. Since the process depended on diffraction of light waves, it required anti-vibration mountings, clean rooms, and highly trained operators. Even at its best, its use of CRTs and film for data storage placed limits on the range depth of images.

At several stages, attaining the frequently over-optimistic expectations for digital computation equipment proved to take far longer than anticipated. For example, the SEASAT system was ready to orbit before its digital processor became available, so a quickly assembled optical recording and processing scheme had to be used to obtain timely confirmation of system operation. In 1978, the first digital SAR processor was developed by the Canadian aerospace company MacDonald Dettwiler (MDA). When its digital processor was finally completed and used, the digital equipment of that time took many hours to create one swath of image from each run of a few seconds of data. Still, while that was a step down in speed, it was a step up in image quality. Modern methods now provide both high speed and high quality.

Although the above specifies the system development contributions of only a few organizations, many other groups had also become players as the value of SAR became more and more apparent. Especially crucial to the organization and funding of the initial long development process was the technical expertise and foresight of a number of both civilian and uniformed project managers in equipment procurement agencies in the federal government, particularly, of course, ones in the armed forces and in the intelligence agencies, and also in some civilian space agencies.

Since a number of publications and Internet sites refer to a young MIT physics graduate named Robert Rines as having invented fine-resolution radar in the 1940s, persons who have been exposed to those may wonder why that has not been mentioned here. Actually, none of his several radar-image-related patents actually had that goal. Instead, they presumed that fine-resolution images of radar object fields could be accomplished by already-known "dielectric lenses", the inventive parts of those patents being ways to convert those microwave-formed images to visible ones. However, that presumption incorrectly implied that such lenses and their images could be of sizes comparable to their optical-wave counterparts, whereas the tremendously larger wavelengths of microwaves would actually require the lenses to have apertures thousands of feet (or meters) wide, like the ones simulated by SARs, and the images would be comparably large. Apparently not only did that inventor fail to recognize that fact, but so also did the patent examiners who approved his several applications, and so also have those who have propagated the erroneous tale so widely. Persons seeking to understand SAR should not be misled by references to those patents.

Relationship to Phased Arrays

A technique closely related to SAR uses an array (referred to as a "phased array") of real antenna elements spatially distributed over either one or two dimensions perpendicular to the radar-range dimension. These physical arrays are truly synthetic ones, indeed being created by synthesis of a collection of subsidiary physical antennas. Their operation need not involve motion relative to targets. All elements of these arrays receive simultaneously in real time, and the signals passing through them can be individually subjected to controlled shifts of the phases of those signals. One result can be to respond most strongly to radiation received from a specific small scene area, focusing on that area to determine its contribution to the total signal received. The coherently detected set of signals received over the entire array aperture can be replicated in several data-processing channels and processed differently in each. The set of responses thus traced to different small scene areas can be displayed together as an image of the scene.

In comparison, a SAR's (commonly) single physical antenna element gathers signals at different positions at different times. When the radar is carried by an aircraft or an orbiting vehicle, those positions are functions of a single variable, distance along the vehicle's path, which is a single mathematical dimension (not necessarily the same as a linear geometric dimension). The signals are stored, thus becoming functions, no longer of time, but of recording locations along that dimension. When the stored signals are read out later and combined with specific phase shifts, the result is the same as if the recorded data had been gathered by an equally long and shaped phased array. What is thus synthesized is a set of signals equivalent to what could have been received simultaneously by such an actual large-aperture (in one dimension) phased array. The SAR simulates (rather than synthesizes) that long one-dimensional phased array. Although the term in the title of this article has thus been incorrectly derived, it is now firmly established by half a century of usage.

While operation of a phased array is readily understood as a completely geometric technique, the fact that a synthetic aperture system gathers its data as it (or its target) moves at some speed means that phases which varied with the distance traveled originally varied with time, hence constituted temporal frequencies. Temporal frequencies being the variables commonly used by radar engineers, their analyses of SAR systems are usually (and very productively) couched in such terms. In particular, the variation of phase during flight over the length of the synthetic aperture is seen as a sequence of Doppler shifts of the received frequency from that of the transmitted frequency. It is significant, though, to realize that, once the received data have been recorded and thus have become timeless, the SAR data-processing situation is also understandable as a special type of phased array, treatable as a completely geometric process.

The core of both the SAR and the phased array techniques is that the distances that radar waves travel to and back from each scene element consist of some integer number of wavelengths plus some fraction of a "final" wavelength. Those fractions cause differences between the phases of the re-radiation received at various SAR or array positions. Coherent detection is needed to capture the signal phase information in addition to the signal amplitude information. That type of detection requires finding the differences between the phases of the received signals and the simultaneous phase of a well-preserved sample of the transmitted illumination.

Every wave scattered from any point in the scene has a circular curvature about that point as a center. Signals from scene points at different ranges therefore arrive at a planar array with different curvatures, resulting in signal phase changes which follow different quadratic variations across a planar phased array. Additional linear variations result from points located in different directions from the center of the array. Fortunately, any one combination of these variations is unique to one scene point, and is calculable. For a SAR, the two-way travel doubles that phase change.

In reading the following two paragraphs, be particularly careful to distinguish between array elements and scene elements. Also remember that each of the latter has, of course, a matching image element.

Comparison of the array-signal phase variation across the array with the total calculated phase variation pattern can reveal the relative portion of the total received signal that came from the only scene point that could be responsible for that pattern. One way to do the comparison is by a correlation computation, multiplying, for each scene element, the received and the calculated field-intensity values array element by array element and then summing the products for each scene element. Alternatively, one could, for each scene element, subtract each array element's calculated phase shift from the actual received phase and then vectorially sum the resulting field-intensity differences over the array. Wherever in the scene the two phases substantially cancel everywhere in the array, the difference vectors being added are in phase, yielding, for that scene point, a maximum value for the sum.

The equivalence of these two methods can be seen by recognizing that multiplication of sinusoids can be done by summing phases which are complex-number exponents of e, the base of natural logarithms.

However it is done, the image-deriving process amounts to "backtracking" the process by which nature previously spread the scene information over the array. In each direction, the process may be viewed as a Fourier transform, which is a type of correlation process. The image-extraction process we use can then be seen as another Fourier transform which is a reversal of the original natural one.

It is important to realize that only those sub-wavelength differences of successive ranges from the transmitting antenna to each target point and back, which govern signal phase, are used to refine the resolution in any geometric dimension. The central direction and the angular width of the illuminating beam do not contribute directly to creating that fine resolution. Instead, they serve only to select the solid-angle region from which usable range data are received. While some distinguishing of the ranges of different scene items can be made from the forms of their sub-wavelength range variations at short ranges, the very large depth of focus that occurs at long ranges usually requires that over-all range differences (larger than a wavelength) be used to define range resolutions comparable to the achievable cross-range resolution.

Data Collection

A model of a German SAR-Lupe reconnaissance satellite inside a Cosmos-3M rocket.

Highly accurate data can be collected by aircraft overflying the terrain in question. In the 1980s, as a prototype for instruments to be flown on the NASA Space Shuttles, NASA operated a synthetic aperture radar on a NASA Convair 990. In 1986, this plane caught fire on takeoff. In 1988, NASA rebuilt a C, L, and P-band SAR to fly on the NASA DC-8 aircraft. Called AIRSAR, it flew missions at sites around the world until 2004. Another such aircraft, the Convair 580, was flown by the Canada Center for Remote Sensing until about 1996 when it was handed over to Environment Canada due to budgetary reasons. Most land-surveying applications are now carried out by satellite observation. Satellites such as ERS-1/2, JERS-1, Envisat ASAR, and RADARSAT-1 were launched explicitly to carry out this sort of observation. Their capabilities differ, particularly in their support for interferometry, but all have collected tremendous amounts of valuable data. The Space Shuttle also carried synthetic aperture radar equipment during the SIR-A and SIR-B missions during the 1980s, the Shuttle Radar Laboratory (SRL) missions in 1994 and the Shuttle Radar Topography Mission in 2000.

The Venera 15 and Venera 16 followed later by the Magellan space probe mapped the surface of Venus over several years using synthetic aperture radar.

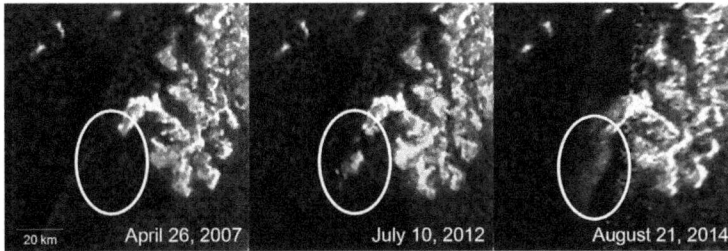

Titan – Evolving feature in Ligeia Mare (SAR; 21 August 2014).

Synthetic aperture radar was first used by NASA on JPL's Seasat oceanographic satellite in 1978 (this mission also carried an altimeter and a scatterometer); it was later developed more extensively on the Spaceborne Imaging Radar (SIR) missions on the space shuttle in 1981, 1984 and 1994. The Cassini mission to Saturn is currently using SAR to map the surface of the planet's major moon Titan, whose surface is partly hidden from direct optical inspection by atmospheric haze. The SHARAD sounding radar on the Mars Reconnaissance Orbiter and MARSIS instrument on Mars Express have observed bedrock beneath the surface of the Mars polar ice and also indicated the likelihood of substantial water ice in the Martian middle latitudes. The Lunar Reconnaissance Orbiter, launched in 2009, carries a SAR instrument called Mini-RF, which was designed largely to look for water ice deposits on the poles of the Moon.

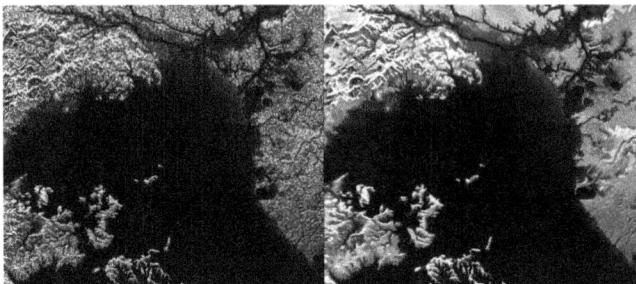

Titan – Ligeia Mare – SAR and clearer despeckled views.

The Mineseeker Project is designing a system for determining whether regions contain landmines based on a blimp carrying ultra-wideband synthetic aperture radar. Initial trials show promise; the radar is able to detect even buried plastic mines.

SAR has been used in radio astronomy for many years to simulate a large radio telescope by combining observations taken from multiple locations using a mobile antenna.

The National Reconnaissance Office maintains a fleet of (now declassified) synthetic aperture radar satellites commonly designated as Lacrosse or Onyx.

In February 2009, the Sentinel R1 surveillance aircraft entered service in the RAF, equipped with the SAR-based Airborne Stand-Off Radar (ASTOR) system.

The German Armed Forces' (Bundeswehr) military SAR-Lupe reconnaissance satellite system has been fully operational since 22 July 2008.

Data Distribution

The Alaska Satellite Facility provides production, archiving and distribution to the scientific community of SAR data products and tools from active and past missions, including the June 2013 release of newly processed, 35-year-old Seasat SAR imagery.

CSTARS downlinks and processes SAR data (as well as other data) from a variety of satellites and supports the University of Miami Rosenstiel School of Marine and Atmospheric Science. CSTARS also supports disaster relief operations, oceanographic and meteorological research, and port and maritime security research projects.

Errors in Radar

Images captured using active remote sensing like radar images will usually be subjected to numerous errors some of which will be discussed here.

Speckle

Microwave signals backscattered from the earth's surface when registered by the sensor can be either in phase or out of phase by varying degrees. The intensity changes will be attributed to atmospheric attenuation effects of scattering, reflection, absorption etc. The phase changes are normally caused due to the target properties. This alteration in the signals will lead to a pattern of brighter and darker pixels in radar images which gives them a grainy appearance, also known as speckle. Speckle is reduced using image processing techniques, such as filtering in the spatial domain, averaging etc. One such technique is multiple look processing, wherein independent images of the same area are generated and then averaged so as to produce a smoother image. If N be the number of statistically independent images being averaged (number of looks), then the amount of speckle will be inversely proportional to the square root of this value. For example, a four look image will have a resolution cell that is four times larger than a one look image. Hence, the major factors contributing to the overall quality of a radar image are the number of looks and the resolution of a system.

Variation in brightness

Radar images especially those from synthetic aperture radar usually contain a gradient in image brightness values in the range direction of the image. There are two reasons for this variation. Firstly, the size of ground resolution cell decreases from near range to far range and secondly, radar backscatter is inversely proportional to the local incident angle. This means that radar images will become darker with increasing range. This will be more pronounced in airborne radar systems with larger range of look angles than the space borne systems. Mathematical models are usually used to correct this effect.

Relief Displacement

Relief displacement is a characteristic feature predominantly seen in side looking radar (SLR) imagery. Whenever a radar pulse encounters a vertical terrain feature, the top of the feature is often reached before the base. Hence, the return signals from the top of a vertical feature will tend to reach the radar antenna before the base of the feature. As a direct consequence, it causes the vertical feature to "lay over" the closer features, thereby making it appear to lean toward the nadir. The effect is termed as layover effect which will be more severe at near range. Another effect commonly observed in radar images is the foreshortening effect. Whenever the slope facing radar antenna is less steep than the line perpendicular to the look direction, layover does not occur. Then the slopes of the surfaces won't be presented in its true size. This effect becomes more predominant as the slope's steepness approaches perpendicularity to the look direction. Sometimes radar imagery will be affected by another characteristic feature known as radar shadow wherein the slopes facing away from the radar antenna will return weak signals or no signal at all.

Radar Remote Sensing from Space

Several satellites have successfully made possible radar remote sensing from space like the experimental spaceborne systems Seasat-1 and the three Shuttle Imaging Radar Systems (SIR-A, SIR-B and SIR-C). Characteristics regarding these instruments are tabulated below. The point to be noted is that radar images acquired using an airborne system will be subjected to large changes in incidence angle across the image swath. Hence, difficulty will be observed in demarcating the backscatter caused by incident angle variations with that from the backscatter actually related to the structure and composition of surface materials/targets in an image. To circumvent this issue, spaceborne systems have only a small change in incident angle enabling easiness in image interpretation. The European space agency (ESA) launched its first remote sensing satellite ERS-1 followed by ERS-2 on July 17, 1991 and April 21, 1995 respectively. Launched in a sun synchronous orbit at an inclination of 98.5^0 and altitude of 785 km, the revisit period of the system is about 16 to18 days. ERS -1 and ERS-2 carry sensors operating in

C, Ku band and an along track scanning radiometer. An advanced polar orbiting Earth observation satellite is Envisat-1, which has an Advanced Synthetic Aperture Radar (ASAR) system onboard it operating in C band. In the image mode, ASAR generates four look high resolution images of 30 m resolution.

Table Characteristics of major Experimental SAR Systems

SAR Systems	Launch Date	Wavelength band	Polarization	Azimuth Resolution (m)	Range Resolution (m)
SEASAT-1	June 27, 1978	L band (23.5 cm)	HH	25	25
SIR-A	November, 1981	L-band	HH	40	40
SIR-B	October 1984	L-band	HH	25	15-45
SIR-C	April 1994 October 1994	X band (X-SAR) C and L bands	HH, HV, VV,VH	25	15-45

The National Space Development Agency of Japan developed the JERS-1 satellite launched on February 11, 1992. It included an L band SAR operating in HH polarization. Similarly, the Canadian remote satellite Radarsat-1 was launched on November 28, 1998 in a sun synchronous orbit at an altitude of 798 km. Radarsat was primarily launched for the purposes of coastal surveillance, land cover mapping, agriculture monitoring etc. However, the near real time monitoring can be extended for disaster management studies like oil spill detection, flood monitoring, landslide identification, identification of soil moisture etc.

Sources of Microwave Data

Globally available microwave data from satellite borne sensors are now made available to end users. An entire list of these data products obtained in microwave and other regions of the electromagnetic spectrum can be obtained from the link.

Two major satellites whose products have been widely used for research purposes are discussed namely that of AMSR-E and TRMM data.

Advanced Microwave Scanning Radiometer (AMSR-E)

AMSRE is a passive microwave radiometer which utilizes 12 channels and 6 different frequencies from 6.9 to 89.0 GHz in the horizontal and vertical polarization. The main objectives of this satellite was to measure geophysical parameters like precipitation, oceanic water vapor, cloud water, near surface wind speed, sea surface temperature, snow cover, sea ice etc. The United States NASA and JAXA of Japan are involved in the algorithm development and implementation for analyzing data from AMSR-E. For each Earth Observing System (EOS) instrument, there exists a suite of data products.

As of october 4, 2011, AMSR-E antenna has stopped working due to aging lubricant in the mechanism. However, the data products along with corresponding metadata and associated documentation are archived and distributed by the Snow and Ice Distributed Active Archive (DAAC) at the National Snow and Ice Data Center. Details regarding the AMSR-E EOS standard data.

Tropical Rainfall Measuring Mission (TRMM)

The TRMM satellite is a joint mission between the National Aeronautics and Space Administration (NASA) of the United States and the National Space Development Agency (NASDA) of Japan. Launched in December 1997 into orbit, the main objectives of TRMM are to measure rainfall and latent heat of condensation exchange of the tropical and subtropical regions of the world. The main instruments onboard TRMM are the TRMM Microwave Imager (TMI), the precipitation radar (PR) and the Visible and Infrared Radiometer System (VIRS). Two additional instruments are also present in TRMM namely the Clouds and Earth's Radiant Energy System (CERES) and the Lightning Imaging System (LIS). More details about the TRMM sensor package can be obtained in Kumerrow et al (1996). Further discussion will be limited to the availability of TRMM microwave products.

The standard TMI data are made available as the first level 1B11 data containing calibrated microwave brightness temperatures captured in 5 frequencies of 10.65, 19.35, 21, 37 and 85.5 GHz in both horizontal and vertical polarizations except for the 21 GHz frequency channel that measures data using just the vertical polarization. The details regarding data available are tabulated below.

Table: TRMM Satellite data products from the active and passive sensors of PR and TMI

1B11	Calibrated TMI (10.65, 19.35, 21, 37 and 85.5 GHz) brightness Temperatures at 5 to 45 km resolution over a 760 km swath
1B21	Calibrated PR (13.8 GHz) power at 4 km horizontal and 250 m vertical Resolutions over a 220 km swath
2A12	TMI Hydrometeor (cloud liquid water, precipitation water, cloud ice, Precipitable ice) profiles in 14 layers at 5 km horizontal resolution, along With latent heat and surface rain, over a 760 km swath.
2A21	PR (13.8 GHz) normalized surface cross section at 4 km horizontal resolution And path attenuation (in case of rain) over a 220 km swath
2A25	PR (13.8 GHz) rain rate, reflectivity and attenuation profiles at 4 km horizontal, and 250 m vertical resolutions over a 220 km swath
3B42	3 hourly 0.25 x 0.25 degree merged TRMM and other satellite estimates
3B31	Rain rate, cloud liquid water, rain water, cloud ice, graupels at 14 levels for a latitude band from 40 degree N to 40 degree S from PR and TMI
3A26	Rain rate probability distribution at surface, 2 km and 4 km for a latitude band from 40 degree N to 40 degree S from PR

As it is not possible to include all the details regarding all the available satellites operating in the microwave spectrum, interested readers can seek more information regarding satellites having active sensors like RISAT-1, RISAT-2, RADARSAT-2, ALOS PALSAR, TERRASAR etc.

Tropical Rainfall Measuring Mission

The Tropical Rainfall Measuring Mission (TRMM) was a joint space mission between NASA and the Japan Aerospace Exploration Agency (JAXA) designed to monitor and study tropical rainfall. The term refers to both the mission itself and the satellite that the mission used to collect data. TRMM was part of NASA's Mission to Planet Earth, a long-term, coordinated research effort to study the Earth as a global system. The satellite was launched on November 27, 1997 from the Tanegashima Space Center in Tanegashima, Japan.

As of July 2014, fuel to maintain orbital altitude was insufficient and NASA ceased station-keeping maneuvers for TRMM, allowing the spacecraft's orbit to slowly decay. Re-entry was originally expected sometime between May 2016 and November 2017. The probe was turned off on April 9, 2015 after its orbital decay accelerated. Re-entry occurred on June 16, 2015 at 06:54 UTC.

Background

Tropical precipitation is a difficult parameter to measure, due to large spatial and temporal variations. However, understanding tropical precipitation is important for weather and climate prediction, as this precipitation contains three-fourths of the energy that drives atmospheric wind circulation. Prior to TRMM, the distribution of rainfall worldwide was known to only a 50% degree of uncertainty.

The concept for TRMM was first proposed in 1984. The science objectives, as first proposed, were:

- To advance understanding of the global energy and water cycles by providing distributions of rainfall and latent heating over the global Tropics.

- To understand the mechanisms through which changes in tropical rainfall influence global circulation and to improve ability to model these processes in order to predict global circulations and rainfall variability at monthly and longer timescales.

- To provide rain and latent heating distributions to improve the initialization of models ranging from 24-hour forecasts to short-range climate variations.

- To help to understand, to diagnose, and to predict the onset and development of the El Niño, Southern Oscillation, and the propagation of the 30- to 60-day oscillations in the Tropics.

- To help to understand the effect that rainfall has on the ocean thermohaline circulations and the structure of the upper ocean.

- To allow cross calibration between TRMM and other sensors with life expectancies beyond that of TRMM itself.

- To evaluate the diurnal variability of tropical rainfall globally.

- To evaluate a space-based system for rainfall measurements.

Japan joined the initial study for the TRMM mission in 1986. Development of the satellite became a joint project between the space agencies of the U.S. and Japan, with Japan providing the Precipitation Radar (PR) and H-II launch vehicle, and the U.S. providing the satellite bus and remaining instruments. The project received formal support from the U.S. congress in 1991, followed by spacecraft construction from 1993 through 1997. TRMM launched from Tanegashima Space Center on 27 November 1997.

Instruments Aboard the TRMM

Precipitation Radar (PR)

The Precipitation Radar was the first space-borne instrument designed to provide three-dimensional maps of storm structure. The measurements yielded information on the intensity and distribution of the rain, on the rain type, on the storm depth and on the height at which the snow melts into rain. The estimates of the heat released into the atmosphere at different heights based on these measurements can be used to improve models of the global atmospheric circulation. The PR operated at 13.8 GHz and measured the 3-d rainfall distribution over land and ocean surfaces. It defined a layer depth of perception and hence measured rainfall that actually reached the latent heat of atmosphere. It had a 4.3 km resolution at radii with 220 km swath.

TRMM Microwave Imager (TMI)

The TRMM Microwave Imager (TMI) was a passive microwave sensor designed to provide quantitative rainfall information over a wide swath under the TRMM satellite. By carefully measuring the minute amounts of microwave energy emitted by the Earth and its atmosphere, TMI was able to quantify the water vapor, the cloud water, and the rainfall intensity in the atmosphere. It was a relatively small instrument that consumed little power. This, combined with the wide swath and the quantitative information regarding rainfall made TMI the "workhorse" of the rain-measuring package on Tropical Rainfall Measuring Mission.

Visible and Infrared Scanner (VIRS)

The Visible and Infrared Scanner was one of the three instruments in the rain-measuring package and serves as a very indirect indicator of rainfall. VIRS, as its name implies, sensed radiation coming up from the Earth in five spectral regions, ranging from visible to infrared, or 0.63 to 12 micrometers. VIRS was included in the primary instrument package for two reasons. First was its ability to delineate rainfall. The second, and even more important reason, was to serve as a transfer standard to other measurements that are made routinely using POES and GOES satellites. The intensity of the radiation in the various spectral regions (or bands) can be used to determine the brightness (visible and near infrared) or temperature (infrared) of the source.

Clouds and the Earth's Radiant Energy Sensor (CERES)

CERES measured the energy at the top of the atmosphere, as well as estimates energy levels within the atmosphere and at the Earth's surface. The CERES instrument was based on the successful Earth Radiation Budget Experiment which used three satellites to provide global energy budget measurements from 1984 to 1993. Using information from very high resolution cloud imaging instruments on the same spacecraft, CERES determines cloud properties, including cloud-amount, altitude, thickness, and the size of the cloud particles. These measurements are important to understanding the Earth's total climate system and improving climate prediction models. It only operated during January - August 1998, and March 2000, so the available data record is quite brief (although later CERES instruments were flown on other missions such as the Earth Observing System (EOS) AM and PM satellites.)

Lightning Imaging Sensor (LIS)

The Lightning Imaging Sensor was a small, highly sophisticated instrument that detects and locates lightning over the tropical region of the globe. The lightning detector was a compact combination of optical and electronic elements including a staring imager capable of locating and detecting lightning within individual storms. The imager's field of view allowed the sensor to observe a point on the Earth or a cloud for 80 seconds, a sufficient time to estimate the flashing rate, which told researchers whether a storm was growing or decaying.

Remote Sensing: Concepts and Technologies

Remote sensing mainly uses technology which helps in achieving a global coverage. The different types of satellites used in orbits are geosynchronous orbit, polar orbit and sun-synchronous orbit. These remote sensing satellites are used to assure that we receive the world's coverage everyday. This section provides a plethora of interdisciplinary topics for better comprehension of the technologies of remote sensing.

Satellites and Orbits

When a satellite is launched into the space, it moves in a well defined path around the Earth, which is called the orbit of the satellite. Gravitational pull of the Earth and the velocity of the satellite are the two basic factors that keep the satellites in any particular orbit. Spatial and temporal coverage of the satellite depends on the orbit. There are three basic types of orbits in use.

- Geo-synchronous orbits

- Polar or near polar orbits

- Sun-synchronous orbits

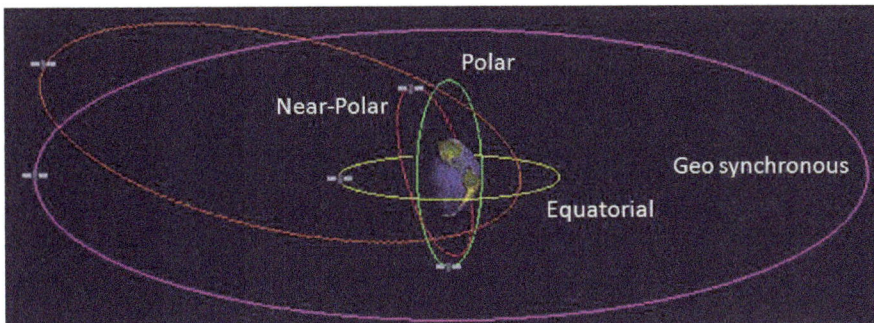

Different types of satellite orbits

Satellite orbits are matched to the capability and objective of the sensor(s) they carry. Orbit selection can vary in terms of altitude (their height above the Earth's surface) and their orientation and rotation relative to the Earth.

Characteristics of Satellite Orbits

The path followed by a satellite in the space is called the orbit of the satellite. Orbits may be circular (or near circular) or elliptical in shape.

Orbital period: Time taken by a satellite to compete one revolution in its orbit around the earth is called orbital period.

It varies from around 100 minutes for a near-polar earth observing satellite to 24 hours for a geo-stationary satellite.

Altitude: Altitude of a satellite is its heights with respect to the surface immediately below it. Depending on the designed purpose of the satellite, the orbit may be located at low (160-2000 km), moderate, and high (~36000km) altitude.

Apogee and perigee: Apogee is the point in the orbit where the satellite is at maximum distance from the Earth. Perigee is the point in the orbit where the satellite is nearest to the Earth as shown in the figure.

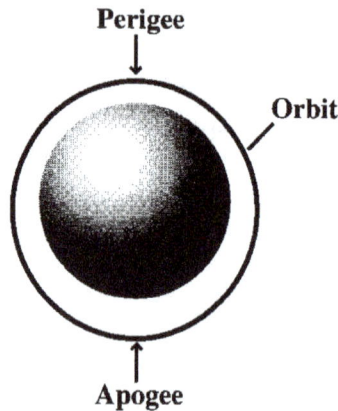

Schematic representation of the satellite orbit showing the Apogee and Perigee

Inclination: Inclination of the orbital plane is measured clockwise from the equator. Orbital inclination for a remote sensing satellite is typically 99 degrees. Inclination of any satellite on the equatorial plane is nearly 180 degrees.

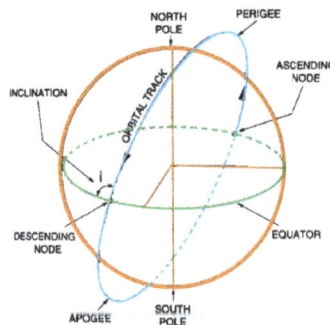

Schematic representation showing the orbital track and inclination

Nadir, ground track and zenith: Nadir is the point of interception on the surface of the Earth of the radial line between the center of the Earth and the satellite. This is the point of shortest distance from the satellite to the earth's surface.

Any point just opposite to the nadir, above the satellite is called zenith.

The circle on the earth's surface described by the nadir point as the satellite revolves is called the ground track. In other words, it is the projection of the satellites orbit on the ground surface.

Swath

Swath of a satellite is the width of the area on the surface of the Earth, which is imaged by the sensor during a single pass.

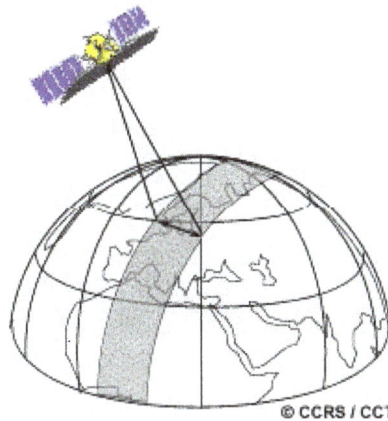

© CCRS / CCT

Schematic representation of the satellite swath

For example, swath width of the IRS-1C LISS-3sensor is 141 km in the visible bands and 148 km in the shortwave infrared band.

Sidelap and Overlap

Overlap is the common area on consecutive images along the flight direction. For example, IRS-1C LISS-3 sensors create 7 km overlap between two successive images.

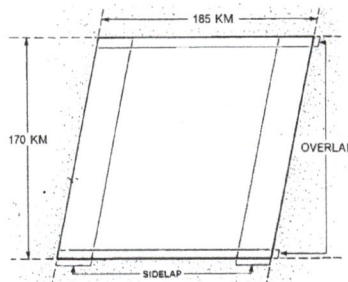

Sidelap and overlap in a typical satellite image

Sidelap is the overlapping areas of the images taken in two adjacent flight lines. For example, sidelap of the IRS 1C LISS-3 sensor at the equator is 23.5 km in the visible bands and 30km in the shortwave infrared band.

As the distance between the successive orbital passes decreases towards the higher latitudes, the sidelap increases. This helps to achieve more frequent coverage of the areas in the higher latitudes. IRS-1C WiFS sensors provide nearly 80-85% overlap and sidelap.

Geosynchronous Orbit

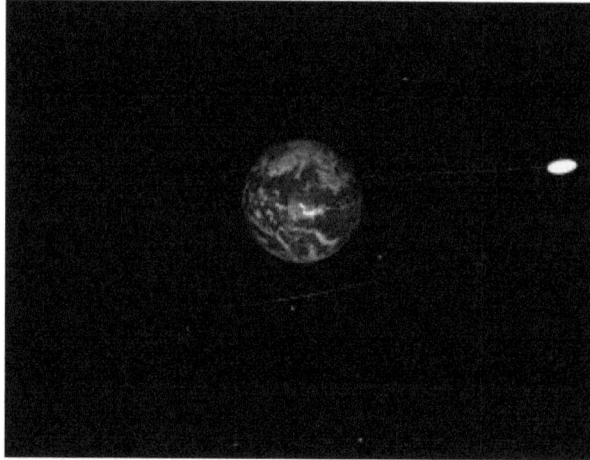

Figure: showing geosynchronous satellite orbiting the Earth.

A geosynchronous orbit (sometimes abbreviated GSO) is an orbit about the Earth of a satellite with an orbital period that matches the rotation of the Earth on its axis (one sidereal day) of approximately 23 hours 56 minutes and 4 seconds. The synchronization of rotation and orbital period means that, for an observer on the surface of the Earth, an object in geosynchronous orbit returns to exactly the same position in the sky after a period of one sidereal day. Over the course of a day, the object's position in the sky traces out a path, whose precise characteristics depend on the orbit's inclination and eccentricity. Satellites are typically launched in an eastward direction. Those that orbit closer to the Earth orbit faster than the Earth rotates and so from the Earth they appear to move in an eastward direction while those that orbit beyond geosynchronous orbit distances will appear to move in a westward direction.

A special case of geosynchronous orbit is the geostationary orbit, which is a circular geosynchronous orbit at zero inclination (that is, directly above the equator). A satellite in a geostationary orbit appears stationary, always at the same point in the sky, to ground observers. Popularly or loosely, the term "geosynchronous" may be used to mean geostationary. Specifically, geosynchronous Earth orbit (GEO) may be a synonym for *geosynchronous equatorial orbit*, or *geostationary Earth orbit*. Communications satellites are often given geostationary orbits, or close to geostationary, so that the satellite antennas that communicate with them do not have to move, but can be pointed permanently at the fixed location in the sky where the satellite appears.

A semi-synchronous orbit has an orbital period of ½ sidereal day, i.e., 11 h 58 min. Relative to the Earth's surface it has twice this period, and hence appears to go around the Earth once every day. Examples include the Molniya orbit and the orbits of the satellites in the Global Positioning System.

Orbital Characteristics

Circular Earth geosynchronous orbits have a radius of 42,164 km (26,199 mi). All Earth geosynchronous orbits, whether circular or elliptical, have the same semi-major axis. In fact, orbits with the same period always share the same semi-major axis:

$$a = \sqrt[3]{\mu \left(\frac{P}{2\pi} \right)^2}$$

where a is the semi-major axis, P is the orbital period, and μ is the geocentric gravitational constant, equal to 398,600.4418 km³/s².

In the special case of a geostationary orbit, the ground track of a satellite is a single point on the equator. In the general case of a geosynchronous orbit with a non-zero inclination or eccentricity, the ground track is a more or less distorted figure-eight, returning to the same places once per sidereal day.

Geostationary Orbit

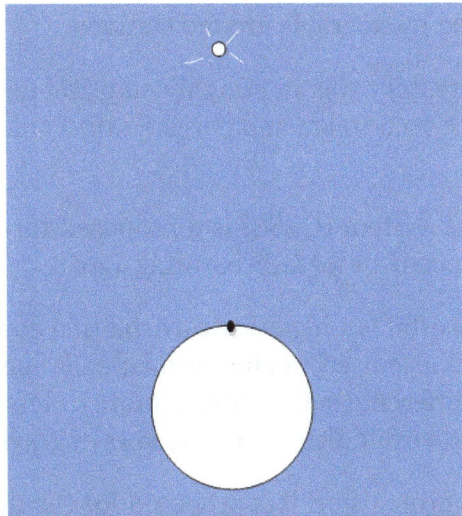

A geostationary satellite above a marked spot on the Equator. An observer on the marked spot will see the satellite remain directly overhead unlike other celestial objects which sweep across the sky.

A geostationary equatorial orbit (GEO) is a circular geosynchronous orbit in the plane of the Earth's equator with a radius of approximately 42,164 km (26,199 mi) (measured from the center of the Earth). A satellite in such an orbit is at an altitude of approximately 35,786 km (22,236 mi) above mean sea level. It maintains the same position relative to the Earth's surface. If one could see a satellite in geostationary orbit, it would appear to hover at the same point in the sky, i.e., not exhibit diurnal motion, while the Sun, Moon, and stars would traverse the skies behind it. The theoretical basis for this novel phenomenon of the sky goes back to Newton's theory of motion and gravity. In that theory, the existence of a geostationary satellite is made possible because the Earth rotates (with respect to an inertial frame in which Newton's laws of motion and gravity hold). However, as a practical device, the geostationary satellite owes much for its realisation to Arthur C. Clarke who proposed it during the 20th century and in whose honour the orbit is called a Clarke orbit. Such orbits are useful for telecommunications satellites.

A perfectly stable geostationary orbit is an ideal that can only be approximated. In practice the satellite drifts out of this orbit because of perturbations such as the solar wind, radiation pressure, variations in the Earth's gravitational field, and the gravitational effect of the Moon and Sun, and thrusters are used to maintain the orbit in a process known as station-keeping.

Other Geosynchronous Orbits

Elliptical geosynchronous orbits can be and are designed for communications satellites in order to keep the satellite within view of its assigned ground stations or receivers. A satellite in an elliptical geosynchronous orbit appears to oscillate in the sky from the viewpoint of a ground station, tracing an analemma in the sky. Satellites in highly elliptical orbits must be tracked by steerable ground stations.

The Infrared Space Observatory was in a highly elliptical geosynchronous orbit with an orbital height of apogee 70,600 km and perigee 1,000 km. It was controlled by two ground stations.

The Quasi-Zenith Satellite System (QZSS) is a proposed three-satellite regional time transfer system and enhancement for GPS covering Japan.

An active geosynchronous orbit is a hypothetical orbit that could be maintained if forces other than gravity were also used, such as a solar sail. Such a statite could be geosynchronous in an orbit different (higher, lower, more or less elliptical, or some other path) from the conic section orbit dictated by the laws of gravity.

A further form of geosynchronous orbit is proposed for the theoretical space elevator, in which one end of the structure is tethered to the ground, maintaining a shorter orbital period than by gravity alone if under tension.

Other related orbit types are:

- Supersynchronous orbit: a disposal / storage orbit above GSO/GEO. Satellites drift in a westerly direction.

- Subsynchronous orbit: a drift orbit close to but below GSO/GEO. Used for satellites undergoing station changes in an eastern direction.

- Graveyard orbit: a supersynchronous orbit where spacecraft are intentionally placed at the end of their operational life.

Other Synchronous Orbits

Syncom 2

Synchronous orbits can only exist for bodies that have a fixed surface (e.g. moons, rocky planets). Without such a surface (e.g. gas giants, black holes) there is no fixed point an orbit can be said to synchronise with. No synchronous orbit will exist if the body rotates so slowly that the orbit would be outside its Hill sphere, or so quickly that it would be inside the body. Large bodies that are held together by gravity cannot rotate that quickly since they would fly apart, so the last condition only applies to small bodies held together by other forces, e.g. smaller asteroids. Most inner moons of planets have synchronous rotation, so their synchronous orbits are, in practice, limited to their leading and trailing (L_4 and L_5) Lagrange points, as well as the L_1 and L_2 Lagrange points, assuming they do not fall within the body of the moon. Objects with chaotic rotations (such as exhibited by Hyperion) are also problematic, as their synchronous orbits change unpredictably.

History

Author Arthur C. Clarke

Author Arthur C. Clarke is credited with proposing the notion of using a geostationary orbit for communications satellites. The orbit is also known as the Clarke Orbit. Together, the collection of artificial satellites in these orbits is known as the Clarke Belt.

The first communications satellite placed in a geosynchronous orbit was Syncom 2, launched in 1963. However, it was in an inclined orbit, still requiring the use of moving antennas. The first communications satellite placed in a geostationary orbit was Syncom 3. Geostationary orbits have been in common use ever since, in particular for satellite television.

Geostationary satellites also carry international telephone traffic but they are being replaced by fiber optic cables in heavily populated areas and along the coasts of less developed regions, because of the greater bandwidth available and lower latency, due to the inherent disconcerting delay in communicating via a satellite in such a high orbit. It takes electromagnetic waves about a quarter of a second to travel from one end to the other end of the link. Thus, two parties talking via satellite are subject to about a half second delay in a round-trip message/response sequence.

Although many populated land locations on the planet now have terrestrial communications facilities (microwave, fiber-optic), even undersea, with more than sufficient capacity, telephone and Internet access is still available only via satellite in many places in Africa, Latin America, and Asia, as well as isolated locations that have no terrestrial facilities, such as Canada's Arctic islands, Antarctica, the far reaches of Alaska and Greenland, and ships at sea.

Polar Orbit

A polar orbit is one in which a satellite passes above or nearly above both poles of the body being orbited (usually a planet such as the Earth, but possibly another body such as the Moon or Sun) on each revolution. It therefore has an inclination of (or very close to) 90 degrees to the equator. A satellite in a polar orbit will pass over the equator at a different longitude on each of its orbits.

Earth Orbits

Polar orbits are often used for earth-mapping, earth observation, capturing the earth as time passes from one point, reconnaissance satellites, as well as for some weather satellites. The Iridium satellite constellation also uses a polar orbit to provide telecommunications services. The disadvantage to this orbit is that no one spot on the Earth's surface can be sensed continuously from a satellite in a polar orbit.

Sun Orbits

Near-polar orbiting satellites commonly choose a Sun-synchronous orbit, meaning that each successive orbital pass occurs at the same local time of day. This can be particularly important for applications such as remote sensing atmospheric temperature, where the most important thing to see may well be *changes* over time which are not aliased onto changes in local time. To keep the same local time on a given pass, the time period of the orbit must be kept as short as possible, this is achieved by keeping the orbit lower to the Earth. However, very low orbits of a few hundred kilometers rapidly decay due to drag from the atmosphere. Commonly used altitudes are between 700 km and 800 km, producing an orbital period of about 100 minutes. The half-orbit on the Sun side then takes only 50 minutes, during which local time of day does not vary greatly.

To retain the Sun-synchronous orbit as the Earth revolves around the Sun during the year, the orbit of the satellite must precess at the same rate, which is not possible if the satellite were to pass directly over the pole. Because of the Earth's equatorial bulge, an orbit inclined at a slight angle is subject to a torque which causes precession; an angle of about 8 degrees from the pole produces the desired precession in a 100-minute orbit.

Sun-synchronous Orbit

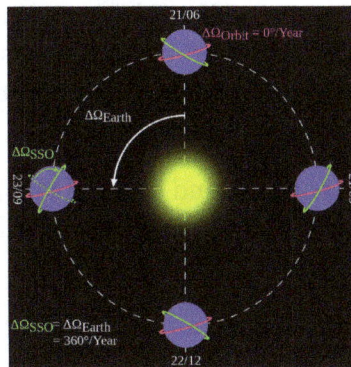

Diagram showing the orientation of a Sun-synchronouses orbit (green) in four points of the year. A non-sun-synchronous orbit (magenta) is also shown for reference. Dates are shown (in white): 21 March (right), 21 June (top), 23 September (left) and 22 December (bottom).

A Sun-synchronous orbit (SSO, also called a heliosynchronous orbit) is a geocentric orbit that combines altitude and inclination in such a way that the satellite passes over any given point of the planet's surface at the same local solar time. Such an orbit can place a satellite in constant sunlight and is useful for imaging, spy, and weather satellites. More technically, it is an orbit arranged in such a way that it precesses once a year. The surface illumination angle will be nearly the same every time that the satellite is overhead. This consistent lighting is a useful characteristic for satellites that image the Earth's surface in visible or infrared wavelengths (e.g. weather and spy satellites) and for other remote sensing satellites (e.g. those carrying ocean and atmospheric remote instruments that require sunlight). For example, a satellite in sun-synchronous orbit might ascend across the equator twelve times a day each time at approximately 15:00 mean local time. This is achieved by having the osculating orbital plane precess (rotate) approximately one degree each day with respect to the celestial sphere, eastward, to keep pace with the Earth's movement around the Sun.

The uniformity of Sun angle is achieved by tuning the inclination to the altitude of the orbit such that the extra mass near the equator causes the orbital plane of the spacecraft to precess with the desired rate: the plane of the orbit is not fixed in space relative to the distant stars, but rotates slowly about the Earth's axis. Typical sun-synchronous orbits are about 600–800 km in altitude, with periods in the 96–100 minute range, and inclinations of around 98° (i.e. slightly retrograde compared to the direction of Earth's rotation: 0° represents an equatorial orbit and 90° represents a polar orbit).

Special cases of the sun-synchronous orbit are the noon/midnight orbit, where the local mean solar time of passage for equatorial longitudes is around noon or midnight, and the dawn/dusk orbit, where the local mean solar time of passage for equatorial longitudes is around sunrise or sunset, so that the satellite rides the terminator between day and night. Riding the terminator is useful for active radar satellites as the satellites' solar panels can always see the Sun, without being shadowed by the Earth. It is also useful for some satellites with passive instruments that need to limit the Sun's influence on the measurements, as it is possible to always point the instruments towards the night side of the Earth. The dawn/dusk orbit has been used for solar observing scientific satellites such as Yohkoh, TRACE, Hinode and PROBA2, affording them a nearly continuous view of the Sun.

Sun-synchronous orbits can happen around other oblate planets, such as Mars. A satellite around the almost spherical Venus, for example, will need an outside push to be in a sun-synchronous orbit.

Technical Details

The angular precession per orbit for an orbit around an oblate planet is (Equation (24) of the article Orbital perturbation analysis (spacecraft)) given by:

$$\Delta\Omega = -2\pi \frac{J_2}{\mu\, p^2} \frac{3}{2} \cos i$$

where

> J_2 is the coefficient for the second zonal term ($1.7555 \cdot 10^{10}$ km^5 / s^2) related to the oblateness of the earth,
>
> μ is the Standard gravitational parameter of the planet (398600.440 km^3 / s^2 for Earth)
>
> p is the semi-latus rectum of the orbit,
>
> i is the inclination of the orbit to the equator.

An orbit will be Sun-synchronous when the precession rate, ρ, equals the mean motion of the Earth about the Sun which is 360° per sidereal year ($1.99096871 \cdot 10^{-7}$ radians / s) so we must set $\Delta\Omega >$ where P is the orbital period.

As the orbital period of a spacecraft is $2\pi\, a\sqrt{\frac{a}{\mu}}$ (where a is the semi-major axis of the orbit) and as $p \approx a$ for a circular or almost circular orbit it follows that

$$\rho \approx -\frac{3 J_2 \cos i}{2 a^{7/2} \mu^{1/2}} = -(360° \text{ per year}) \times (a/12352 \text{ km})^{-7/2} \cos i = -(360° \text{ per year}) \times (P/3.795 \text{ hrs})^{-7/3} \cos i$$

or when ρ is 360° per year,

$$\cos i \approx -\frac{\rho\sqrt{\mu}}{\frac{3}{2} J_2} a^{\frac{7}{2}} = -(a/12352 \text{ km})^{7/2} = -(P/3.795 \text{ hrs})^{7/3},$$

As an example, for a=7200 km (the spacecraft about 800 km over the Earth surface) one gets with this formula a Sun-synchronous inclination of 98.696 deg.

Note that according to this approximation cos i equals −1 when the semi-major axis equals 12 352 km, which means that only smaller orbits can be Sun-synchronous. The period can be in the range from 88 minutes for a very low orbit (a=6554 km, i=96°) to 3.8 hours (a=12 352 km, but this orbit would be equatorial with i=180°). (A period longer than 3.8 hours may be possible by using an eccentric orbit with p<12 352 km but a>12 352 km.)

If one wants a satellite to fly over some given spot on Earth every day at the same hour, it can do between 7 and 16 orbits per day, as shown in the following table. (The table has been calculated assuming the periods given. The orbital period that should be used is actually slightly longer. For instance, a retrograde equatorial orbit that passes over the same spot after 24 hours has a true period about 365/364 ≈ 1.0027 times longer than the time between overpasses. For non-equatorial orbits the factor is closer to 1.)

Orbits per day	Period (hrs)	Height above Earth's surface (km)	Maximum latitude
16	= 1 hr 30 min	282	83.4°
15	= 1 hr 36 min	574	82.3°
14	≈ 1 hr 43 min	901	81.0°
13	≈ 1 hr 51 min	1269	79.3°
12		1688	77.0°
11	≈ 2 hrs 11 min	2169	74.0°
10	= 2 hrs 24 min	2730	69.9°
9	= 2 hrs 40 min	3392	64.0°
8		4189	54.7°
7	≈ 3 hrs 26 min	5172	37.9°

When one says that a Sun-synchronous orbit goes over a spot on the earth at the same local time each time, this refers to mean solar time, not to apparent solar time. The Sun will not be in exactly the same position in the sky during the course of the year.

The Sun-synchronous orbit is mostly selected for Earth observation satellites that should be operated at a relatively constant altitude suitable for its Earth observation instruments, this altitude typically being between 600 km and 1000 km over the Earth surface. Because of the deviations of the gravitational field of the Earth from that of a homogeneous sphere that are quite significant at such relatively low altitudes a strictly circular orbit is not possible for these satellites. Very often a frozen orbit is therefore selected that is slightly higher over the Southern hemisphere than over the Northern hemisphere. ERS-1, ERS-2 and Envisat of European Space Agency as well as the MetOp spacecraft of EUMETSAT are all operated in Sun-synchronous, "frozen" orbits.

Spatial and Spectral Resolution

A digital image consists of an array of pixels. Each pixel contains information about a small area on the land surface, which is considered as a single object.

Spatial resolution is a measure of the area or size of the smallest dimension on the Earth's surface over which an independent measurement can be made by the sensor.

It is expressed by the size of the pixel on the ground in meters.

Angular Resolution

Angular resolution or spatial resolution describes the ability of any image-forming device such as an optical or radio telescope, a microscope, a camera, or an eye, to distinguish small details of an object, thereby making it a major determinant of image resolution.

Definition of Terms

Resolving power is the ability of an imaging device to separate (i.e., to see as distinct) points of an object that are located at a small angular distance or it is the power of an optical instrument to separate far away objects, that are close together, into individual images. The term *resolution* or *minimum resolvable distance* is the minimum distance between distinguishable objects in an image, although the term is loosely used by many users of microscopes and telescopes to describe resolving power. In scientific analysis, in general, the term "resolution" is used to describe the precision with which any instrument measures and records (in an image or spectrum) any variable in the specimen or sample under study.

Explanation

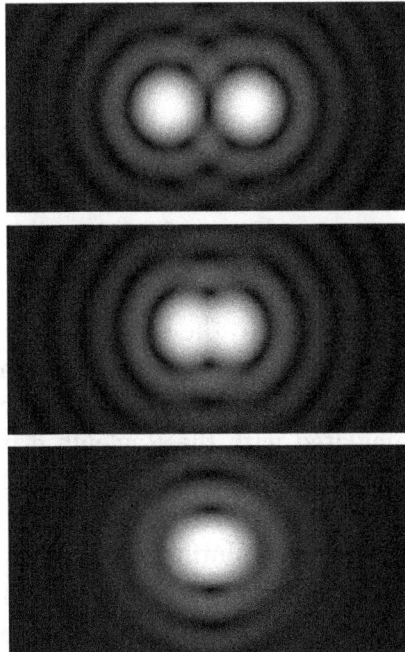

Airy diffraction patterns generated by light from two points passing through a circular aperture, such as the pupil of the eye. Points far apart (top) or meeting the Rayleigh criterion (middle) can be distinguished. Points closer than the Rayleigh criterion (bottom) are difficult to distinguish.

The imaging system's resolution can be limited either by aberration or by diffraction causing blurring of the image. These two phenomena have different origins and are unrelated. Aberrations can be explained by geometrical optics and can in principle be solved by increasing the optical quality — and consequently the cost — of the system. On the other hand, diffraction comes from the wave nature of light and is determined by the finite aperture of the optical elements. The lens' circular aperture is analogous to a two-dimensional version of the single-slit experiment. Light passing through the lens interferes with itself creating a ring-shape diffraction pattern, known as the Airy pattern, if the wavefront of the transmitted light is taken to be spherical or plane over the exit aperture.

The interplay between diffraction and aberration can be characterised by the point spread function (PSF). The narrower the aperture of a lens the more likely the PSF is dominated by diffraction. In that case, the angular resolution of an optical system can be estimated (from the diameter of the aperture and the wavelength of the light) by the Rayleigh criterion defined by Lord Rayleigh: two point sources are regarded as just resolved when the principal diffraction maximum of one image coincides with the first minimum of the other. If the distance is greater, the two points are well resolved and if it is smaller, they are regarded as not resolved. Rayleigh defended this criteria on sources of equal strength.

Considering diffraction through a circular aperture, this translates into:

$$\theta = 1.220 \frac{\lambda}{D}$$

where θ is the *angular resolution* (radians), λ is the wavelength of light, and D is the diameter of the lens' aperture. The factor 1.220 is derived from a calculation of the position of the first dark circular ring surrounding the central Airy disc of the diffraction pattern. This number is more precisely 1.21966989... (A245461), the first zero of the order-one Bessel function of the first kind $J_1(x)$ divided by π.

The formal Rayleigh criterion is close to the empirical resolution limit found earlier by the English astronomer W. R. Dawes who tested human observers on close binary stars of equal brightness. The result, $\theta = 4.56/D$, with D in inches and θ in arcseconds is slightly narrower than calculated with the Rayleigh criterion: A calculation using Airy discs as point spread function shows that at Dawes' limit there is a 5% dip between the two maxima, whereas at Rayleigh's criterion there is a 26.3% dip. Modern image processing techniques including deconvolution of the point spread function allow resolution of binaries with even less angular separation.

The angular resolution may be converted into a *spatial resolution*, $\Delta\ell$, by multiplication of the angle (in radians) with the distance to the object. For a microscope, that distance is close to the focal length f of the objective. For this case, the Rayleigh criterion reads:

$$\Delta\ell = 1.220 \frac{f\lambda}{D}.$$

This is the size, in the imaging plane, of smallest object that the lens can resolve, and also the radius of the smallest spot to which a collimated beam of light can be focused. The size is proportional to wavelength, λ, and thus, for example, blue light can be focused to a smaller spot than red light. If the lens is focusing a beam of light with a finite extent (e.g., a laser beam), the value of D corresponds to the diameter of the light beam, not the lens. Since the spatial resolution is inversely proportional to D, this leads to the slightly surprising result that a wide beam of light may be focused to a smaller spot than a narrow one. This result is related to the Fourier properties of a lens.

A similar result holds for a small sensor imaging a subject at infinity: The angular resolution can be converted to a spatial resolution on the sensor by using f as the distance to the image sensor; this relates the spatial resolution of the image to the f-number, $f/\#$:

$$\Delta\ell \approx 1.220\frac{f\lambda}{D} = 1.22\lambda \cdot (f/\#)$$

Since this is the radius of the Airy disk, the resolution is better estimated by the diameter, $2.44\lambda \cdot (f/\#)$

Specific Cases

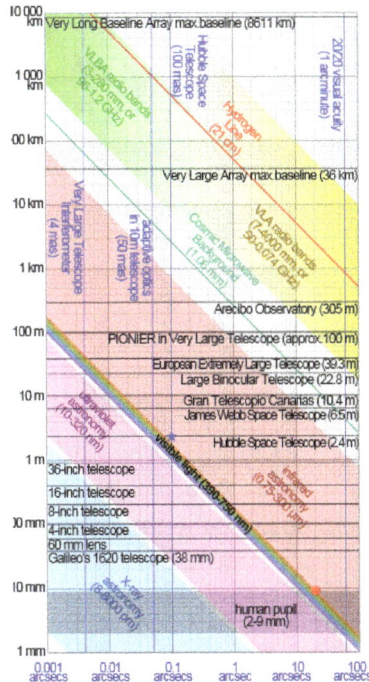

Log-log plot of aperture diameter vs angular resolution at the diffraction limit for various light wavelengths compared with various astronomical instruments. For example, the blue star shows that the Hubble Space Telescope is almost diffraction-limited in the visible spectrum at 0.1 arcsecs, whereas the red circle shows that the human eye should have a resolving power of 20 arcsecs in theory, though normally only 60 arcsecs.

Single Telescope

Point-like sources separated by an angle smaller than the angular resolution cannot be resolved. A single optical telescope may have an angular resolution less than one arcsecond, but astronomical seeing and other atmospheric effects make attaining this very hard.

The angular resolution R of a telescope can usually be approximated by

$$R = \frac{\lambda}{D}$$

where λ is the wavelength of the observed radiation, and D is the diameter of the telescope's objective. The Resulting R is in radians. For example, in the case of yellow light with a wavelength of 580 nm, for a resolution of 0.1 arc second, we need D=1.2 m. Sources larger than the angular resolution are called extended sources or diffuse sources, and smaller sources are called point sources.

This formula, for light with a wavelength of about 562 nm, is also called the Dawes' limit.

Telescope Array

The highest angular resolutions can be achieved by arrays of telescopes called astronomical interferometers: These instruments can achieve angular resolutions of 0.001 arcsecond at optical wavelengths, and much higher resolutions at radio wavelengths. In order to perform aperture synthesis imaging, a large number of telescopes are required laid out in a 2-dimensional arrangement.

The angular resolution R of an interferometer array can usually be approximated by

$$R = \frac{\lambda}{B}$$

where λ is the wavelength of the observed radiation, and B is the length of the maximum physical separation of the telescopes in the array, called the baseline. The resulting R is in radians. Sources larger than the angular resolution are called extended sources or diffuse sources, and smaller sources are called point sources.

For example, in order to form an image in yellow light with a wavelength of 580 nm, for a resolution of 1 milli-arcsecond, we need telescopes laid out in an array that is 120 m × 120 m.

Microscope

The resolution R (here measured as a distance, not to be confused with the angular resolution of a previous subsection) depends on the angular aperture :

$$R = \frac{1.22\lambda}{NA_{condenser} + NA_{objective}} \quad \text{where} \quad NA = \eta \sin\theta.$$

Here NA is the numerical aperture, θ is half the included angle α of the lens, which depends on the diameter of the lens and its focal length, η is the refractive index of the medium between the lens and the specimen, and λ is the wavelength of light illuminating or emanating from (in the case of fluorescence microscopy) the sample.

It follows that the NAs of both the objective and the condenser should be as high as possible for maximum resolution. In the case that both NAs are the same, the equation may be reduced to:

$$R = \frac{0.61\lambda}{NA} \approx \frac{\lambda}{2NA}$$

The practical limit for θ is about 70°. In a dry objective or condenser, this gives a maximum NA of 0.95. In a high-resolution oil immersion lens, the maximum NA is typically 1.45, when using immersion oil with a refractive index of 1.52. Due to these limitations, the resolution limit of a light microscope using visible light is about 200 nm. Given that the shortest wavelength of visible light is violet ($\lambda \approx$ 400 nm),

$$R = \frac{1.22 \times 400\text{nm}}{1.45 + 0.95} = 203\text{nm}$$

which is near 200 nm.

Oil immersion objectives can have practical difficulties due to their shallow depth of field and extremely short working distance, which calls for the use of very thin (0.17 mm) cover slips, or, in an inverted microscope, thin glass-bottomed Petri dishes.

However, resolution below this theoretical limit can be achieved using optical near-fields (Near-field scanning optical microscope) or a diffraction technique called 4Pi STED microscopy. Objects as small as 30 nm have been resolved with both techniques. In addition to this Photoactivated localization microscopy can resolve structures of that size, but is also able to give information in z-direction (3D).

Notes

In the case of laser beams, a Gaussian Optics analysis is more appropriate than the Rayleigh criterion, and may reveal a smaller diffraction-limited spot size than that indicated by the formula above.

Spectral Resolution

The spectral resolution of a spectrograph, or, more generally, of a frequency spectrum, is a measure of its ability to resolve features in the electromagnetic spectrum. It is usually denoted by $\Delta\lambda$, and is closely related to the resolving power of the spectrograph, defined as

$$R = \frac{\lambda}{\Delta\lambda},$$

where $\Delta\lambda$ is the smallest difference in wavelengths that can be distinguished at a wavelength of λ. For example, the Space Telescope Imaging Spectrograph (STIS) can distinguish features 0.17 nm apart at a wavelength of 1000 nm, giving it a resolution of 0.17 nm and a resolving power of about 5,900. An example of a high resolution spectrograph is the *Cryogenic High-Resolution IR Echelle Spectrograph* (CRIRES) installed at ESO's Very Large Telescope, which has a spectral resolving power of up to 100,000.

Doppler Effect

The spectral resolution can also be expressed in terms of physical quantities, such as velocity; then it describes the difference between velocities Δv that can be distinguished through the Doppler effect. Then, the resolution is Δv and the resolving power is

$$R = \frac{c}{\Delta v}$$

where c is the speed of light. The STIS example above then has a spectral resolution of 51 km/s.

IUPAC Definition

IUPAC defines resolution in optical spectroscopy as the minimum wavenumber, wavelength or frequency difference between two lines in a spectrum that can be distinguished. Resolving power, R, is given by the transition wavenumber, wavelength or frequency, divided by the resolution.

Temporal and Radiometric Resolutions

Temporal Resolution

Temporal resolution describes the number of times an object is sampled or how often data are obtained for the same area.

The absolute temporal resolution of a remote sensing system to image the same area at the same viewing angle a second time is equal to the repeat cycle of a satellite.

The repeat cycle of a near polar orbiting satellite is usually several days, eg., for IRS-1C and Resourcesat-2 it is 24 days, and for Landsat it is 18 days. However due to the off-nadir viewing capabilities of the sensors and the sidelap of the satellite swaths in the adjacent orbits the actual revisit period is in general less than the repeat cycle.

The actual temporal resolution of a sensor therefore depends on a variety of factors, including the satellite/sensor capabilities, the swath overlap, and latitude.

Because of some degree of overlap in the imaging swaths of the adjacent orbits, more frequent imaging of some of the areas is possible. The following figure shows the schematic of the image swath sidelap in a typical near polar orbital satellite.

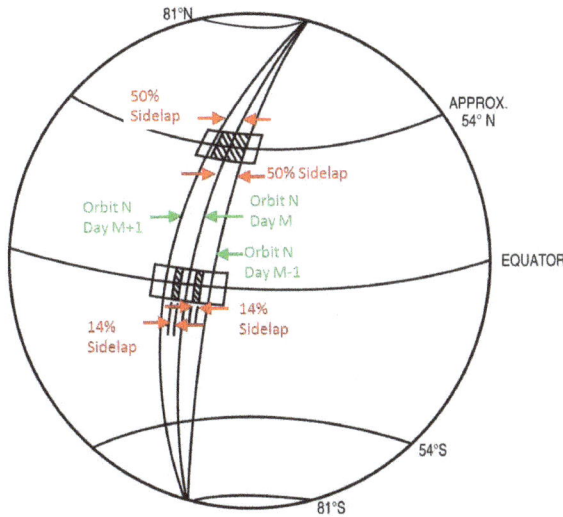

Sidelap in a typical near polar satellite orbit

In the figure above, it can be seen that the sidelap increases with latitude. Towards the polar region, satellite orbits come closer to each other compared to the equatorial regions. Therefore for the polar region the sidelap is more. Therefore more frequent images are available for the polar region. The figure below shows the path of a typical near-polar satellite.

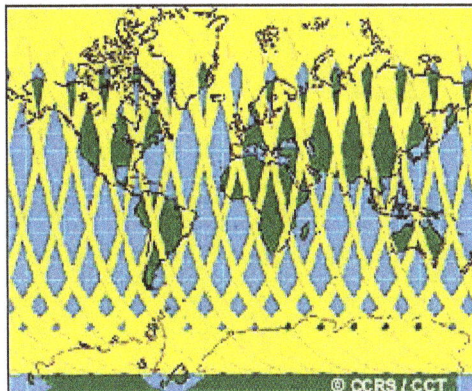

Orbit of a typical near-polar satellite

In addition to the sidelap, more frequent imaging of any particular area of interest is achieved in some of the satellites by pointing their sensors to image the area of interest between different satellite passes. This is referred as the off-nadir viewing capability.

For example: using pointable optics, sampling frequency as high as once in 1-3 days are achieved for IKONOS, whereas the repeat cycle of the satellite is 14 days.

Images of the same area of the Earth's surface at different periods of time show the variation in the spectral characteristics of different features or areas over time. Such multi-temporal data is essential for the following studies.

- Land use/ land cove classification

- Temporal variation in land use / land cover

- Monitoring of a dynamic event like

    - Cyclone

    - Flood

    - Volcano

    - Earthquake

Flood studies: Satellite images before and after the flood event help to identify the aerial extent of the flood during the progress and recession of a flood event. The Great Flood of 1993 or otherwise known as the Great Mississippi and Missouri Rivers Flood of 1993, occurred from April and October 1993 along the Mississippi and Missouri rivers and their tributaries. The flood was devastating affecting around $15 billion and was one of the worst such disasters occurring in United States. The following figure shows the landsat TM images taken during a normal period and during the great flood of 1993. Comparison of the two images helps to identify the inundated areas during the flood.

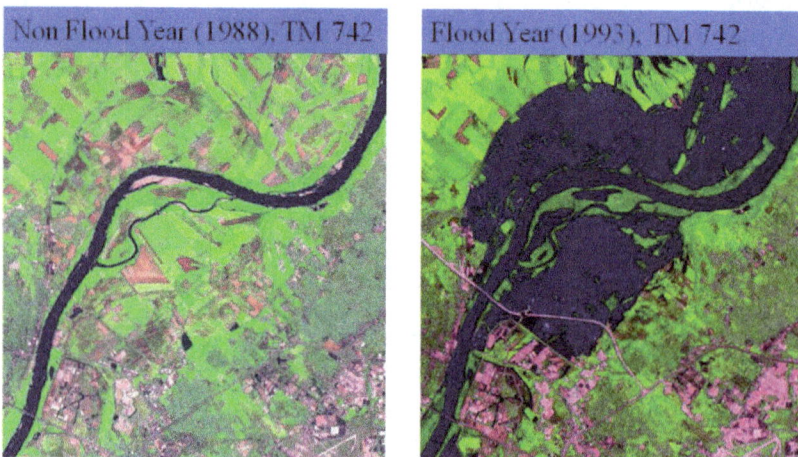

Landsat TM images of the Mississipi River during non-flood period and during the great flood of 1993

Land use/ land cover classification: Temporal variation in the spectral signature is valuable in land use/ land cover classification. Comparing multi-temporal images, the presence of features over time can be identified, and this is widely adopted for classifying various types of crops / vegetation. For example, during the growing season, the vegetation characteristics change continuously. Using multi-temporal images it is possible to monitor such changes and thus the crop duration and crop growth stage can be identified, which can be used to classify the crop types viz., perennial crops, long or short duration crops.

The following figure shows the MODIS data product for the Krishna River Basin in different months in 2001. Images of different months of the year help to differentiate the forest areas, perennial crops and short duration crops.

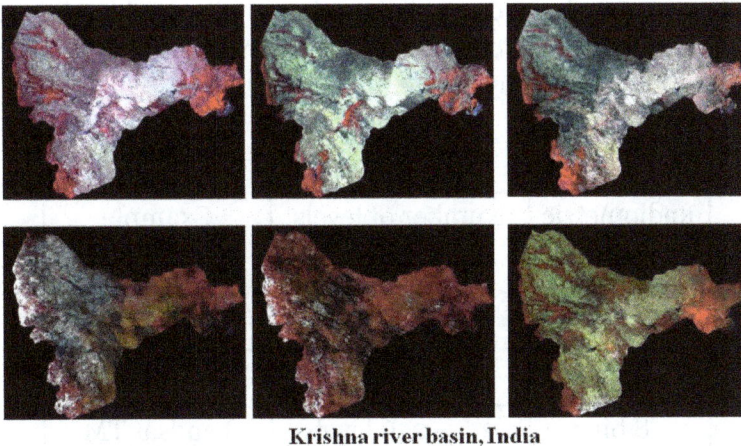

Krishna river basin, India

FCC (RGB): 2,1,6 (NIR, red, MIR1)

False Color Composites (FCC) of the Krishna River Basin generated from the MODIS data for different months in 2001.

The figure represents False Color Composites (FCC) of the river basin.

In remote sensing the term resolution is used to represent the resolving power, which includes not only the capability to identify the presence of two objects, but also their properties. In qualitative terms the resolution is the amount of details that can be observed in an image. Four types of resolutions are defined for the remote sensing systems.

- Spatial resolution

- Spectral resolution

- Temporal resolution

- Radiometric resolution

Radiometric Resolution

Radiometric resolution of a sensor is a measure of how many grey levels are measured between pure black (no reflectance) to pure white. In other words, radiometric resolution represents the sensitivity of the sensor to the magnitude of the electromagnetic energy.

The finer the radiometric resolution of a sensor the more sensitive it is to detecting small differences in reflected or emitted energy or in other words the system can measure more number of grey levels.

Radiometric resolution is measured in bits.

Each bit records an exponent of power 2 (e.g. 1 bit = 2^1 = 2). The maximum number of brightness levels available depends on the number of bits used in representing the recorded energy. For example, the table below shows the radiometric resolution and the corresponding brightness levels available.

Table: Radiometric resolution and the corresponding brightness levels.

Radiometric resolution	Number of levels	Example
1 bit	2^1 – 2 levels	
7 bit	2^7 – 128 levels	IRS 1A & 1B
8 bit	2^8 – 256 levels	Landsat TM
11 bit	2^{11} – 2048 levels	NOAA-AVHRR

Thus, if a sensor used 11 bits to record the data, there would be 2^{11}=2048 digital values available, ranging from 0 to 2047. However, if only 8 bits were used, then only 2^8=256 values ranging from 0 to 255 would be available. Thus, the radiometric resolution would be much less.

Image data are generally displayed in a range of grey tones, with black representing a digital number of 0 and white representing the maximum value (for example, 255 in 8-bit data). By comparing a 2-bit image with an 8-bit image, we can see that there is a large difference in the level of detail discernible depending on their radiometric resolutions. In an 8 bit system, black is measured as 0 and white is measured as 255. The variation between black to white is scaled into 256 classes ranging from 0 to 255. Similarly, 2048 levels are used in an 11 bit system as shown in the figure.

Finer the radiometric resolution, more the number of grey levels that the system can record and hence more details can be captured in the image.

Variation in the brightness levels recorded at different radiometric resolution
(Source: Gibson 2000)

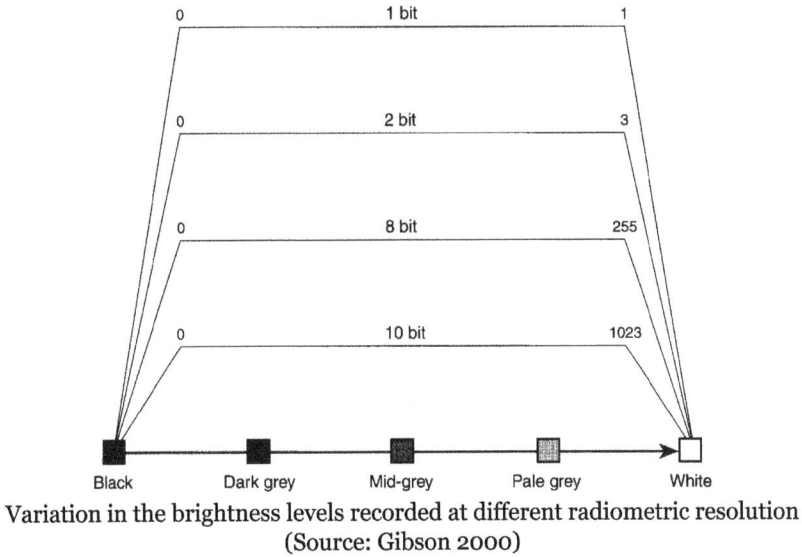

The figure below shows the comparison of a 2-bit image (coarse resolution) with an 8-bit image (fine resolution), from which a large difference in the level of details is apparent depending on their radiometric resolutions.

As radiometric resolution increases, the degree of details and precision available will also increase. However, increased radiometric resolution may increase the data storage requirements.

2 Bit Data (Coarse)   8 Bit Data (Fine)

Comparison of a coarse resolution 2-bit image with a fine resolution 8-bit image

In an image, the energy received is recoded and represented using Digital Number (DN). The DN in an image may vary from 0 to a maximum value, depending up on the number of gray levels that the system can identify i.e., the radiometric resolution. Thus, in addition to the energy received, the DN for any pixel varies with the radiometric resolution. For the same amount of energy received, in a coarse resolution image (that can record less number of energy level) a lower value is assigned to the pixel compared to a fine resolution image (that can record more number of energy level). This is explained with the help of an example below.

Example: A RS system with a radiometric resolution of 6 bits assigns a DN of 28, 45 and 48 to three surfaces. What would be the equivalent DNs for the same surfaces if the measurements were taken with a 3 bit system?

The DNs recorded by the 3-bit system range from 0 to 7 and this range is equivalent to 0-63 for the 6 bit system.

0	1	2	3	4	5	6	7	(3 bit)
0	9	18	27	36	45	54	63	(6 bit)

Therefore a DN of 28 on the 6-bit system will be recorded as 3 in the 3-bit system. A 6-bit system could record the difference in the energy at levels 45 and 47, whereas in a 3-bit system both will be recorded as 5.

Therefore when two images are to be compared, they must be of same radiometric resolution.

Multispectral Remote Sensing

Multi-band imaging employs the selective sensing of the energy reflected in multiple wavelength bands in the range 0.3 to 0.9 µm. Generally broad bands are used in multi-band imaging. Multi-spectral scanners operate using the same principle, however using more number of narrower bands in a wider range varying from 0.3 to approximately 14 µm. Thus multi-spectral scanners operate in visible, near infrared (NIR), mid-infrared (MIR) and thermal infrared regions of the electro-magnetic radiation (EMR) spectrum.

Thermal scanners are special types of multi-spectral scanners that operate only in the thermal portion of the EMR spectrum. Hyperspectral sensing is the recent development in the multi- spectral scanning, where hundreds of very narrow, contiguous spectral bands of the visible, NIR, MIR portions of the EMR spectrum are employed.

This chapter gives a brief description of the multispectral remote sensing. Different types of multispectral scanners and their operation principles are covered in this chapter.

Multispectral Scanners

A Multispectral scanner (MSS) simultaneously acquires images in multiple bands of the EMR spectrum. It is the most commonly used scanning system in remote sensing.

For example the MSS onboard the first five Landsat missions were operational in 4 bands: 0.5-0.6, 0.6-0.7, 0.7-0.8, 0.8-1.1 μm. Similarly, IRS LISS-III sensors operate in four bands (0.52-0.59, 0.62-0.68, 0.77-0.86, 1.55-1.70 μm) three in the visible and NIR regions and one in the MIR region of the EMR spectrum.

Spectral reflectance of the features differs in different wavelength bands. Features are identified from the image by comparing their responses over different distinct spectral bands. Broad classes, such as water and vegetation, can be easily separated using very broad wavelength ranges like visible and near-infrared. However, for more specific classes viz.,

vegetation type, rock classification etc, much finer wavelength ranges and hence finer spectral resolution are required.

The following figures shows the bands 4, 5, 6 and 7 obtained from Lansdat1 MSS and the standard FCC.

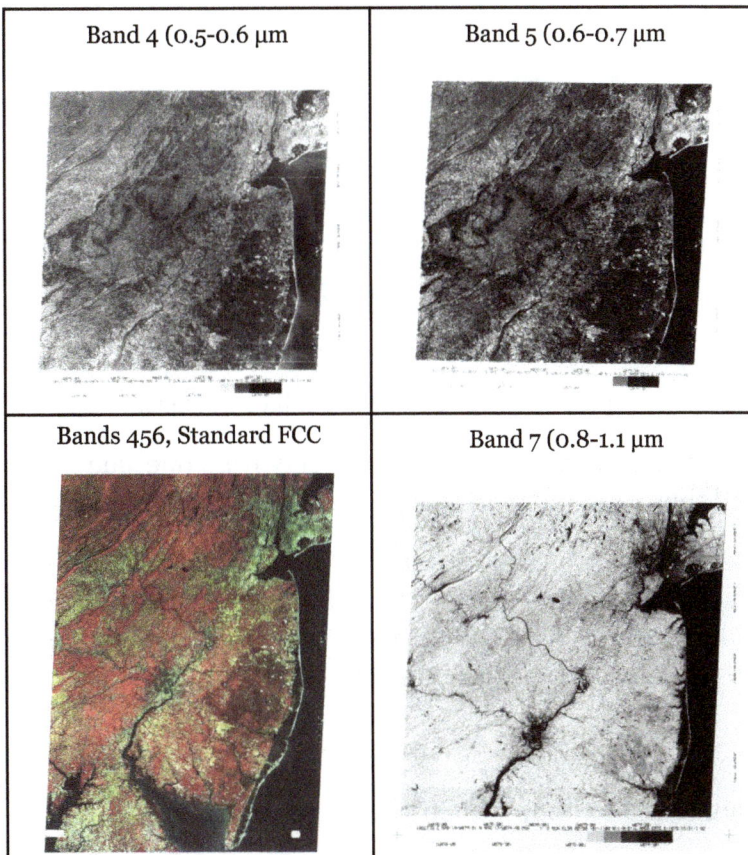

Landsat-1 MSS images of an area obtained in different spectral bands and the standard FCC

The figure clearly displays how water, vegetation and other features are displayed in different bands, and how the combination of different bands helps the feature identification.

Airborne or space-borne MSS systems generate two-dimensional images of the terrain beneath the aircraft. Two different approaches are adopted for this: Across-track (whiskbroom) scanning and Along-track (push broom) scanning.

Across-track Scanning

Across-track scanner is also known as whisk-broom scanner. In across track scanner, rotating or oscillating mirrors are used to scan the terrain in a series of lines, called scan lines, which are at right angles to the flight line. As the aircraft or the platform moves forward, successive lines are scanned giving a series of contiguous narrow strips. Schematic representation of the operational principle of a whisk-broom scanner is shown in the figure.

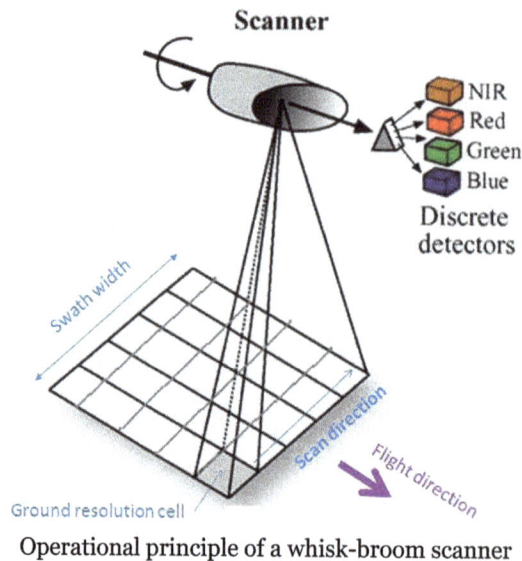

Operational principle of a whisk-broom scanner

The scanner thus continuously measures the energy from one side to the other side of the platform and thus a two-dimensional image is generated.

The incoming reflected or emitted radiation is separated into several thermal and non-thermal wavelength components using a dichroic grating and a prism. An array of electro-optical detectors, each having peak spectral sensitivity in a specific wavelength band, is used to measure each wavelength band separately.

Along-track Scanning

Along-track scanner is also known as push-broom scanner.

Along-track scanners also use the forward motion of the platform to record successive scan lines and build up a two-dimensional image, perpendicular to the flight direction. However, along-track scanner does not use any scanning mirrors, instead a linear array of detectors is used to simultaneously record the energy received from multiple ground resolution cells along the scan line. This linear array typically consists of numerous charged coupled devices (CCDs). A single array may contain more than 10,000 individual detectors. Each detector element is dedicated to record the energy in a single column as shown in the figure. Also, for each spectral band, a separate linear array of detectors is used. The arrays of detectors are arranged in the focal plane of the scanner in such a way that the each scan line is viewed simultaneously by all the arrays. The array of detectors are pushed along the flight direction to scan the successive scan lines, and hence the name push-broom scanner. A two- dimensional image is created by recording successive scan lines as the aircraft moves forward.

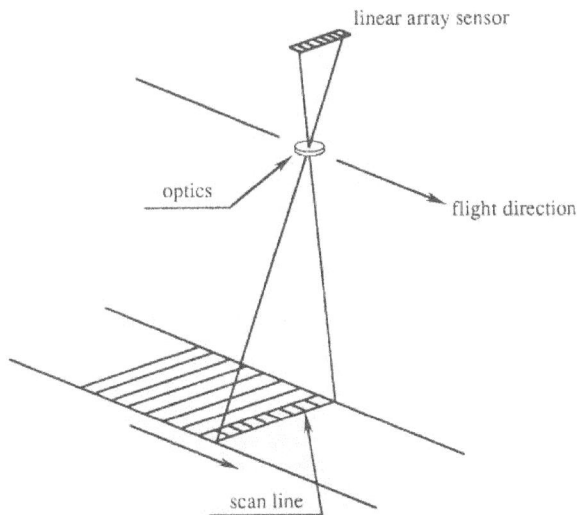

Schematic representation of a Push-Broom Scanner

The linear array of detectors provides longer dwell time over each ground resolution cell, which increases the signal strength. This also increases the radiometric resolution. In a push- broom scanner, size of the ground resolution cell is determined by the IFOV of a single detector. Thus, finer spatial and spectral resolution can be achieved without impacting radiometric resolution.

Thematic Mapper

Thematic Mapper (TM) is an advanced multispectral scanner designed to achieve higher spatial, spectral and radiometric accuracy. It was introduced by NASA during the Landsat-4 mission. The TM used in the Landsat mission was operational in 7 bands. These bands are more refined compared to the MSS and are designated for some potential application. Principal applications of each of the Landsat TM bands are shown in Table below.

Landsat TM spectral bands and their potential applications

Band	Spectral range (μm)	Principal application
1	0.45-0.52	Coastal water mapping, soil-vegetation differentiation, deciduous-coniferous differentiation
2	0.52-0.6	Green reflectance by healthy vegetation
3	0.63-0.69	Chlorophyl absorption for plant species differentiation
4	0.76-0.90	Biomass surveys, water body delineation
5	1.55-1.72	Vegetation moisture measurement, snow-cloud differentiation
6	10.4-12.5	Plant heat stress measurement, other thermal mapping
7	2.08-2.35	Hydrothermal mapping

Thermal Scanner

Thermal scanner is a special kind of across track multispectral scanner which senses the energy in the thermal wavelength range of the EMR spectrum. Thermal infrared radiation refers to electromagnetic waves with wavelength 3-14 μm. The atmosphere absorbs much of the energy in the wavelength ranging from 5-8 μm. Due to the atmospheric effects, thermal scanners are generally restricted to 3-5 μm and 8-14 μm wavelength ranges.

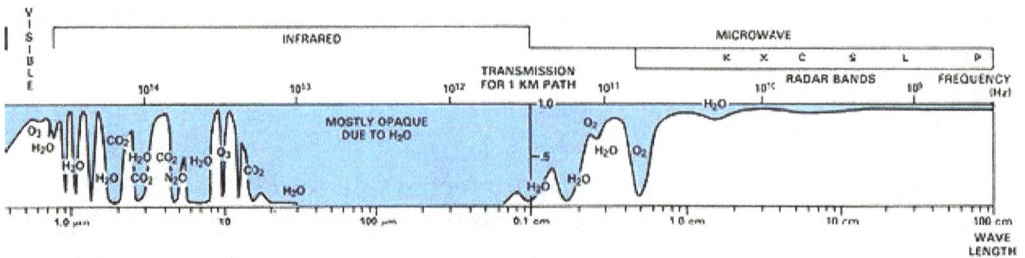

Figure: Shows a day time thermal image if the San-Francisco region recorded using 8.5-13.0 μm thermal wavelength region. The runway of the airport appears in light tone as the thermal emission from the runway is more in the day time.

Day time thermal image (8.5-13 µm) of San Francisco

The Advanced Spaceborne Thermal Emission and Reflection Radiometer (ASTER) onboard Terra, TIMS developed jointly by NASA JPL and the Daedalus Corporation are some of the examples. ASTER data is used to create detailed maps of land surface temperature, reflectance, and elevation. TIMS is used as an airborne geologic remote sensing tool to acquire mineral signatures to discriminate minerals like silicate and carbonate. It uses 6 wavelength channels as shown in the table below.

Spectral bands of the TIMS

Channel	Wavelength µm
1	8.2-8.6
2	8.6-9.0
3	9.0-9.4
4	9.4-10.2
5	10.2-11.2
6	11.2-12.2

Since the energy received at the sensor decreases as the wavelength increases, larger IFOVs are generally used in thermal sensors to ensure that enough energy reaches the detector for a reliable measurement. Therefore the spatial resolution of thermal sensors is usually fairly coarse, relative to the spatial resolution possible in the visible and reflected infrared. However, due to the relatively long wavelength, atmospheric scattering is minimal in thermal scanning. Also since the reflected solar radiation is not measured in thermal scanning, it can be operated in both day and night times.

Principle Involved in the Thermal Sensing

In thermal scanning the energy radiated from the land surface is measured using thermal sensors. The thermal emission is the portion of internal energy of surface that is transformed into radiation energy.

Radiation from a Blackbody and Real Materials

A blackbody is a hypothetical, ideal radiator that totally absorbs and re-emits all energy incident upon it. Emissivity (ε) is the factor used to represent the radiant exitance of a material compared to that of a blackbody. Thus

$$\varepsilon = \frac{\text{radiant exitance of an object at a given temperature}}{\text{radiant exitance of a blackbody at thesame temperature}}$$

An ideal blackbody (the body which transforms all internal energy into radiation energy) has emissivity equal to 1. The emissivity of real surfaces ranges from 0 to 1.

Emissivity of a material varies with the wavelength, viewing angle and temperature. If the emissivity of a material varies with wavelength, it is called a selective radiator. If a material has constant emissivity, which is less than 1, in all the wavelengths it is called a grey body.

In the thermal scanning, the radiant energy from the surface is measured.

According to the Stefan-Bolzmann law, the radiant exitance (M) from a black body is given by

$$M = \sigma T^4$$

Where, σ is the Stefan-Boltzmann's constant = 5.6697×10^{-8} W m^{-2} K^{-4}, T is the temperature of the black body (K)

For a real material, it can be extended as

$$M = \varepsilon \sigma T^4$$

The thermal sensors record the radiant energy M from the surface. Thus if we know the emissivity ε, we can determine the real surface temperature. But in general, in satellite remote sensing the target features are unknown and hence are their emissivities. In such cases, the brightness temperature of the surface is determined, which is the surface temperature if that were a blackbody.

Thermal Imaging

For the thermal energy sensing, typically quantum or photon detectors containing electrical charge carriers are used. The principle behind the thermal scanning is the direct relationship between the photons of radiation falling on the detector and the energy levels of the electrical charge carriers.

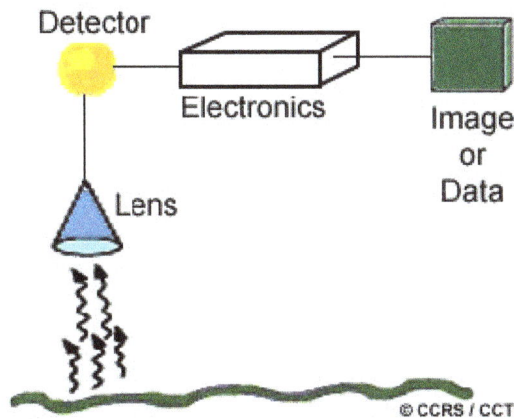

Schematic representation of a thermal sensor operational principle

Some of the commonly used detectors are mercury-doped germanium (sensitive in the range 3-14 μm), indium antimonide (sensitive in the region 3-5 μm) and mercury cadmium telluride (sensitive in the region 8-14 μm).

A thermal scanner image, also known as thermogram, is a pictorial representation of the detector's response on a line-by-line basis. Thus areas having higher radiant/ brightness temperature are displayed as lighter tomes in the image. Most of the thermal scanning operations in geologic and water resources studies are qualitative in nature, wherein only the relative difference in the radiant temperature are obtained.

Information about the temperature extremes, heating and the cooling rates are used to interpret the type and condition of the object, For example, water reaches maximum temperature slower than rocks or soils and therefore, terrain temperatures are normally higher than the water temperature during the day time and lower during the night.

Some of the important applications of thermal remote sensing image are the following.

- Geological studies: determining rock type and structures
- Soil mapping

- Soil moisture studies

- Study of evapotranspiration in vegetation

- Detection of heat looses in buildings

- Detection of damages of steam pipelines and caliducts

- Detection of subsurface fires(e.g. coal seams)

Hyperspectral Sensors

Hyperspectral sensors (also known as imaging spectrometers) are instruments that acquire images in several, narrow, contiguous spectral bands in the visible, NIR, MIR, and thermal infrared regions of the EMR spectrum. Hyperspectral sensors may be along-track or across- track.

A typical hyperspectral scanner records more than 100 bands and thus enables the construction of a continuous reflectance spectrum for each pixel.

For example, the Hyperion sensor onboard NASA's EO-1 satellite images the earth's surface in 220 contiguous spectral bands, covering the region from 400 nm to 2.5 μm, at a ground resolution of 30 m. The AVIRIS sensor developed by the JPL contains four spectrometers with a total of 224 individual CCD detectors (channels), each with a spectral resolution of 10 nanometers and a spatial resolution of 20 meters.

The following figure shows the schematic representation of the hyperspectral imaging process.

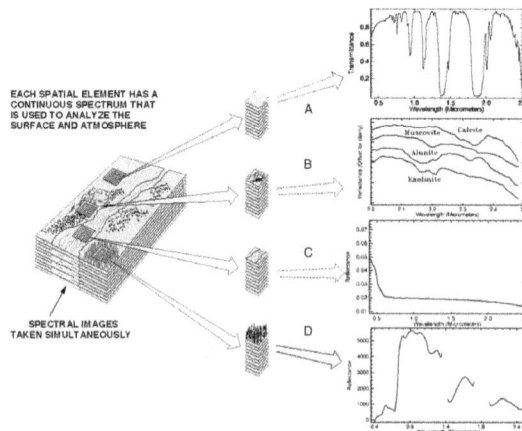

Schematic representation of the hyperspectral imaging (Source: Kruse, 2012)

From the data acquired in multiple, contiguous bands, the spectral curve for any pixel can be calculated that may correspond to an extended ground feature.

Depending on whether the pixel is a pure feature class or the composition of more than one feature class, the resulting plot will be either a definitive curve of a "pure" feature or a composite curve containing contributions from the several features present. Spectral curves of the pixels are compared with the existing spectral library to identify the targets. All pixels whose spectra match the target spectrum to a specified level of confidence are marked as potential targets.

Hyperspectral AVIRIS image of the San Juan Valley of Colorado is shown below. The figure below shows the spectral curves for different crop classes generated using the reflectance from multiple bands of the AVIRIS image. Spectral curves generated from the image are used to identify the vegetation or crop type in the circular fields and are verified with the ground data.

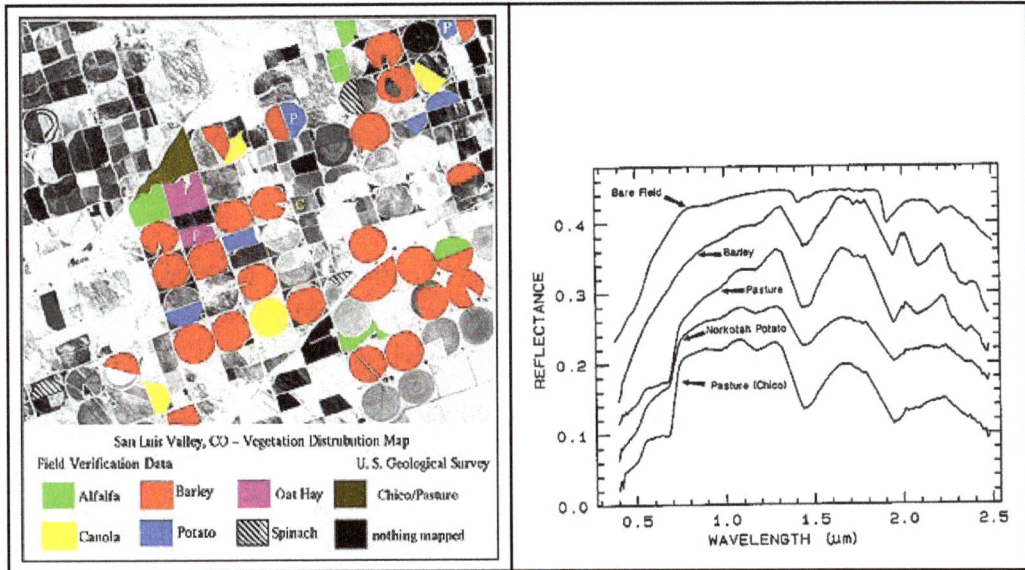

Hyperspectral AVIRIS image of the San Juan Valley of Colorado and the spectral signature curves generated for different fields

Hyperspectral imaging has wide ranging applications in mining, geology, forestry, agriculture, and environmental management.

Features of Remote Sensing Satellite

There are many characteristics that describe any satellite remote sensing systems. Satellite's orbit (including its altitude, period, inclination and the equatorial crossing time), repeat cycle, spatial resolution, spectral characteristics, radiometric properties are a few of them.

Landsat Satellite Program

Landsat is the longest running program for acquiring satellite imageries of the Earth.

First satellite in the series, Landsat-1 was launched in July 1972. It was a collaborative effort of NASA and the US department of the Interior. The program was earlier called Earth Resources Technology Satellites (ERTSs) and was later on renamed as Landsat in 1975. The mission consists of 8 satellites launched successively. The recent one in the series Landsat-8, which is also called Landsat Data Continuity Mission (LDCM) was launched in February, 2013.

The following figure shows the time line of the Landsat satellite program.

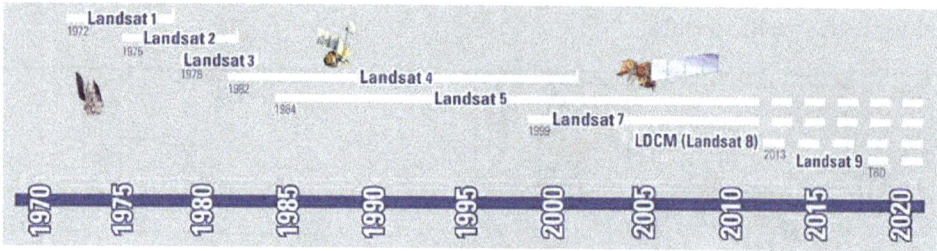

Time line of the Landsat satellite program

Different types of sensors viz., Return Beam Vidicom (RBV), Multispectral Scanner (MSS), Thematic Mapper, Enhanced Thematic Mapper (ETM), and Enhanced Thematic Mapper Plus (ETM+) have been used in various Landsat missions.

Landsat missions use sun-synchronous, near polar orbits at different altitudes for each mission

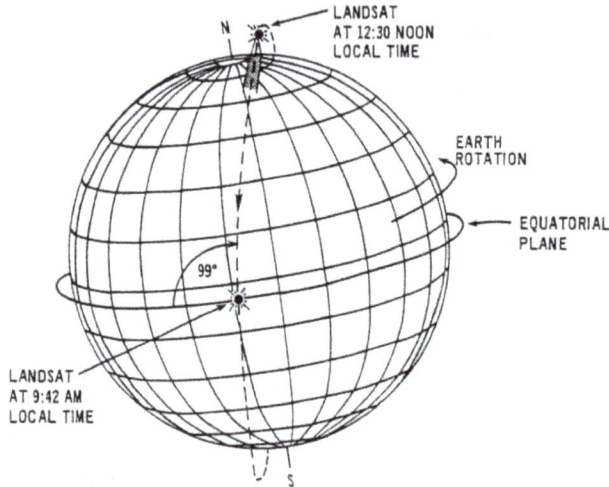

Source: http://landsat.gsfc.nasa.gov/

Typical orbit of a satellite in the Landsat program

The following table gives the details of different Landsat missions including the type of sensors , spatial, temporal and radiometric resolution.

Table: Details of the orbit and sensors of different Landsat missions

Mission	Landsat-1		Landsat-2		Landsat-3		Landsat-4		Landsat-5		Landsat-6	Landsat-7	Landsat-8(L-CDM)	
Mission period	1972-1978		1975-1982		1978-1983		1982-2001		1984-2012		1993, failed	April 1999 -	Feb 2013 -	
Orbit	Sun-synchronous, near-polar													
Altitude	917 km		917 km		917km		705 km		706 km			705km	705 km	
Inclination	99.2 deg		99.2 deg		99.2 deg		98.2 deg		98.2 deg			98.2deg	98.2 deg 10	
Eq. crossing	9:30am		9:30am		9:30am		9:45am		9:45am			10am	am	
Period (min) (+/- 15min)	103.34		103		103		99					98.9	98.9	
No. orbits / day	14		14		14		14		14			14	14	
Repeat cycle	18		18		18		16		16			16	16	
Swath width	185		185		185		185		185			185	185	
Sensors	RBV	MSS	RBV	MSS	RBV	MSS	MSS	TM	MSS	TM	ETM	ETM+	OLI	TIRS
Bands	1-3	4-7	1-3	4-7	1-4	4-8	1-4	1-7	1-4	1-7	1-8	1-8	1-9	1-2
Spatial	80	82	80	82	80	82	79	30	79	30	B1-B5,B7: 30	B1-B5,B7: 30	30	100
resolution (m)						B8:240		B6:120		B6:120	B6: 120	B6: 60	B8:15	
											B8: 15	B8: 15		
Radiometric	6	B1-B3:7	6	B1-B3:7	6	B1-B3:7	B1-B3:7	8	B1-B3:7	8	8	8	12	12
resolution		B4: 6		B4: 6		B4: 6	B4: 6		B4: 6					

Landsat satellites typically complete 14 orbits in a day. The following figure shows the orbital path of a Landsat satellite.

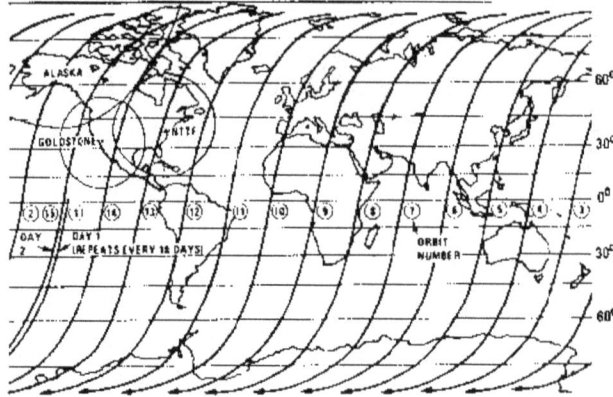

Successive orbits of a typical Landsat satellite

Landsat 4 and 5 maintained 8 days out of phase, so that when both were operational, 8-day repeat coverage could be maintained. MSS used in the Landsat programs employs across line scanning to generate two-dimensional image.

Table: Spectral bands of the OLI and TIPS sensors of the Landsat-8 mission.

Operational Land Imager (OLI)			Thermal Infrared Scanner (TIRS)		
Band	Wavelength (μm)	Remark	Band	Wavelength (μm)	Remark
1	0.43-0.45	Coastal aerosol detection	1	10.60-11.19	Thermal infrared
2	0.45-0.51	Blue	2	11.50-12.51	Thermal infrared
3	0.53-0.59	Green			
4	0.64-0.67	Red			
5	0.85-0.88	Near infrared			
6	1.57-1.65	Short wave infrared			
7	2.11-2.29	Short wave infrared			
8	0.50-0.68	Panchromatic			
9	1.36-1.38	Cirrus cloud detection			

SPOT Satellite Program

SPOT (*Systeme Pour l'Observation de la Terre*) was designed by the Centre National d'Etudes Spatiales (CNES), France as a commercially oriented earth observation program. The first satellite of the mission, SPOT-1 was launched in February, 1986. This was the first earth observation satellite that used a linear array of sensors and the push broom scanning techniques. Also these were the first system to have pointable/steerable optics, enabling side- to-side off-nadir viewing capabilities.

The figure below shows the timeline of various missions in the SPOT satellite program

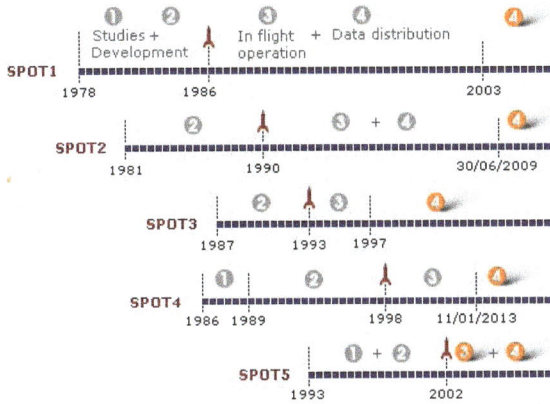

The recent satellite in the SPOT program, SPOT 6 was launched on September 2012.

SPOT 1, 2 and 3 carried two identical High Resolution Visible (HRV) imaging systems. Each HRVs were capable of operating either in the panchromatic mode or in the MSS mode. HRVs used along-track, push-broom scanning methods. Each HRV contained four CCD sub- arrays. A 6000-element sub-array was used for recording in the panchromatic mode and the remaining 3 arrays, each with 3000 elements, were used for the MSS mode. Due to the off- nadir viewing capability, HRV was also used for stereoscopic imaging. Frequency with which the stereoscopic coverage can be obtained varies with the latitude; more frequent imaging is possible near the polar region compared to the equatorial region.

SPOT 4 carried the High Resolution Visible and Infrared (HRVIR) sensor and the vegetation instrument (VI). HRVIR also includes two identical sensors, both together capable of giving 120km swath width at nadir.

SPOT-5 carries two high resolution geometric (HRG) instruments, a single high resolution stereoscopic (HRS) instrument, and a vegetation instrument (VI). Details of the sensors used in various SPOT 4 and 5 missions are summarized in the table.

Table: Details of the sensors used in SPOT 4 and SPOT 5 missions

SPOT-4				SPOT-5			
HRVIR		VI		HRG		HRS and VI	
Bands	Wavelength (μm)	Bands	Wavelength (μm)	Bands	Wavelength (μm)	Bands	Wavelength (μm)
1	0.53-0.59	1	0.43-0.47	PAN	0.48-0.71	PAN	0.49-0.69
2	0.61-0.68	2	0.61-0.68	1	0.50-0.59	1	0.45-0.52
3	0.79-0.89	3	0.79-0.89	2	0.61-0.68	2	0.61-0.58
4	1.58-1.75	4	1.58-1.75	3	0.78-0.89	3	0.78-0.89
				4	1.58-1.75	4	1.58-1.75

SPOT-6 mission employs two New AstroSat Optical Modular Instruments (NAOMI). The instrument operates in 5 spectral bands, including one panchromatic band. Details of these bands are given in the table.

Table: Spectral bands of the NAOMI used in SPOT-6 mission

Band	Wavelength (μm)	Remark
PAN	0.45-0.745	Panchromatic
1	0.450-0.525	Blue
2	0.530-0.590	Green
3	0.625-0.695	Red
4	0.760-0.890	Near infrared

The following table gives the details of various SPOT missions. Mission period, orbit characteristics, sensors employed, and the resolution details are given in the table.

Table: Details of the SPOT satellite missions

Mission	SPOT-1 SPOT-2 SPOT-3	SPOT-4		SPOT-5			SPOT-6
Mission period	1986-2003 1990-2009 1993-1997	1998-2013		2002-			2012-
Orbit	Sun-synchronous, near-polar, circular						
Altitude Inclination Eq. crossing Period	822 98.7 10:30 AM 101.4	822 98.7 10:30 AM 101.4		822 98.7 10:30 AM 101.4			694 98.2 10:30 AM 98.79
Repeat cycle	26 days (More frequent revisit is achieved due to the off-nadir viewing capability)						
Sensors	HRV	HRVIR	VI	HRG	HRS	VI	NAOMI
Bands Spatial resolution Radiometric resolution	PAN and B1-B3 PAN:10m , MSS:20m 8bit	B1-B4 B1-PAN: 10m B1-B4 MSS: 20m 8bit	B0 B2-B4 1000 10 bit	PAN B1-B4 PAN:2.5-5m MSS: 10m B4: 20m 8 bit	PAN 10m	B0 B2-B4 1000 10 bit	PAN B1-B4 PAN: 2m MSS: 8m 12 bit

Pointable optics used in the program enables off-nadir viewing. This increases the frequency of viewing viz., 7 additional viewings at equator and 11 additional viewings at 45deg latitude. Due to the off-nadir viewing capabilities, stereo imaging is also possible. Stereo pairs, used for relief perception and elevation plotting (Digital Elevation Modelling), are formed from two SPOT images.

IRS Satellite Program

Indian Remote Sensing (IRS) satellite system is one of the largest civilian remote sensing satellite constellations in the world used for earth observation. Objective of the program is to provide a long-term space-borne operational capability for the observation and management of the natural resources. IRS satellite data have been widely used in studies related to agriculture, hydrology, geology, drought and flood monitoring, marine studies and land use analyses.

The first satellite of the mission IRS-1A was launched in 1988. IRS satellites orbit the Earth in sun-synchronous, near-polar orbits at low altitude. Various missions in the IRS satellite program employ various sensors viz., LISS-1, LISS-2, LISS-3, WiFS, AWiFS etc.

Spectral bands used in various sensors of the IRS satellite program are given in the table.

Table: Spectral bands used in various sensors of the IRS satellites.

Sensor	LISS-1 and 2	LISS-3	LISS-4	WiFS	AWiFS
Wavelength bands (µm)	0.45-0.52 0.52-0.59 0.62-0.68 0.77-0.86	0.52-0.59 0.62-0.68 0.77-0.86 1.55-1.70	0.52-0.59 0.62-0.68 0.77-0.86	0.62-0.68 0.77-0.86	0.52-0.59 0.62-0.68 0.77-0.86 1.55-1.70

Details of various satellite missions of the IRS program, including the mission period, orbit characteristics, sensors and resolutions are given in the table.

Table: Details of the various satellites of the IRS satellite program.

Satellite	IRS-1A	IRS-1B	IRS-1C	IRS-1D	IRS-P2	Cartosat-2	Resourcesat-2
Period	1988-1996	1991-2003	1995-2007	1997-2010	2003-	2007-	2011-
Orbit	Sun-synchronous, Polar						
Eq. crossing	10:30am						
Altitude	904		817		817	630	822
Inclination	99.08		98.6		98.7	97.91	98.73

Repeat cycle (days)	22		24 LISS-4 and AW-iFS : 5 Revisit: 4		24	310	24
Sensors	LISS-1, LISS-2A and 2B		PAN, LISS-3, WiFS		LISS-3 and 4, AWiFS	PAN camera	LISS-3 and 4, AWiFS
Bands	B1-B4		PAN, LISS-3 B1-B4 WiFS B1-B2		LISS-3 B1-B4 LISS-4 B1-B3 AwiFS B1-B4	PAN (0.5-0.85μm)	LISS-3 B1-B4 LISS-4 B1-B3 AwiFS B1-B4
Spatial resolution	72.5m	36.25m	PAN:5.8m LISS-3: 23m (B4:70m)		LISS-3:23.5 LISS-4: 5.8 AWiFS: 56m	0.81m	LISS-3:23.5 LISS-4: 5.8 AWiFS: 56m
Radio-metric reso-lution (Bits)	7	7	7	7	LISS-3 and 4: 7 AwiFS: 10	10	LISS-3 and 4: 10 AwiFS: 12

IRS-P6 LISS-IV multispectral mode image shows the centre of Marseille, France, in natural colours

Parts of Paris as viewed by Cartosat-2 in 2011

Parts of Himalayas as viewed by the AWiFS sensor

Ikonos

IKONOS is a commercial Earth observation satellite, and was the first to collect publicly available high-resolution imagery at 1- and 4-meter resolution. It offers multispectral (MS) and panchromatic (PAN) imagery. The IKONOS launch was called "one of the most significant developments in the history of the space age". IKONOS imagery began being sold on 1 January 2000.

It derived its name from the Greek term *eikōn* for image.

History

IKONOS was originated under the Lockheed Corporation as the Commercial Remote Sensing System (CRSS) satellite. In April 1994 Lockheed was granted one of the first licenses from the U.S. Department of Commerce for commercial satellite high-resolution imagery. On 25 October 1995 partner company Space Imaging received a license from the Federal Communications Commission (FCC) to transmit telemetry from the satellite in the eight-gigahertz Earth Exploration Satellite Services band. Prior to launch, Space Imaging changed the name of the satellite system to *IKONOS*. The name comes from the Greek word for "image".

Two satellites were originally planned for operation. *IKONOS-1* was launched on 27 April 1999 at 18:22 UTC from Vandenberg Air Force Base SLC-6, but Athena II rocket's payload fairing did not separate due to an electrical malfunction, resulting in the satellite failing to reach orbit and falling into the atmosphere over the South Pacific Ocean.

IKONOS-2 was built in parallel with and as an identical twin to *IKONOS-1*. Completion of its construction was projected for July 1999 with a January 2000 launch. In reaction to the loss of *IKONOS-1*, the spacecraft was renamed *IKONOS* and its processing accelerated, resulting in a launch on 24 September 1999 at 18:22 UTC, also from Vandenberg aboard an Athena II rocket. *IKONOS* has a mass of 817 kilograms (1,800 lb) and operates in a Sun-synchronous, near-polar, circular 681 km (423 mi) orbit. It has five imaging sensors, one panchromatic and four multispectral (blue, green, red, and near-infrared), and has a nadir image swath width of 11.3 km (7 mi).

In December 2000, *IKONOS* received the "Best of What's New" Grant Award in Aviation & Space from *Popular Science* magazine. The acquisition of Space Imaging and its assets by Orbimage was announced in September 2005 and finalized in January 2006. The merged company was renamed GeoEye, which was itself acquired by DigitalGlobe in January 2013. DigitalGlobe operated *IKONOS* until its retirement on 31 March 2015.

Specifications

Spacecraft

IKONOS is a three-axis stabilized spacecraft designed by Lockheed Martin. The design later became known as the LM900 satellite bus system. The satellite's altitude is measured by two star trackers and a sun sensor and controlled by four reaction wheels; location knowledge is provided by a GPS receiver. The design life is seven years; S/C body size=1.83 m × 1.57 m (hexagonal configuration); S/C mass = 817 kg; power = 1.5 kW provided by three solar panels.

The LM900 spacecraft is a three-axis stabilized bus that is designed to carry scientific payloads in LEOs. It provides precision pointing on an ultra stable highly agile platform. Payloads for a variety of scientific and remote sensing applications may be accommodated including laser sensors, imagers, radar sensors, electro-optical and astronomical sensors, as well as planetary sensors. The LM900 bus shares a hardware heritage with Iridium, which is the basis for the LM700 bus.

Communications

IKONOS conducts telemetry, tracking and control in the 8345.968–8346.032 MHz band (downlink) and 2025–2110 MHz band (uplink). Downlink data carrier operates in the 8025-8345 MHz band.

Optics & Detectors

IKONOS has a primary mirror aperture of 0.7 m (2.3 feet), and a folded optical focal length of 10 m (about 33 feet) using 5 mirrors. The main mirror features a honeycomb design to reduce mass. The detectors at the focal plane include a pan-chromatic and a multi-spectral sensor, with 13500 pixels and 3375 pixels respectively (cross-track). Total instrument mass is 171 kg (377 pounds) and it uses 350 watts.

Imaging Capabilities

Spatial Resolution

- 0.8 m panchromatic (1-m PAN)
- 4-meter multispectral (4-m MS)
- 1-meter pan-sharpened (1-m PS)

Spectral Resolution

Band	1-m PAN	4-m MS & 1-m PS
1 (Blue)	0.45–0.90 μm	0.445–0.516 μm
2 (Green)	*	0.506–0.595 μm
3 (Red)	*	0.632–0.698 μm
4 (Near IR)	*	0.757–0.853 μm

Temporal Resolution

The revisit rate for IKONOS is three to five days off-nadir and 144 days for true-nadir.

Radiometric Resolution

The sensor collects data with an 11-bit (0–2047) sensitivity and are delivered in an unsigned 16-bit (0–65535) data format. From time-to-time the data are rescaled down to 8-bit (0–255) to decrease file size. When this occurs much of the sensitivity of the data needed by remote sensing scientists is lost.

Swath

11 km × 11 km (single scene)

QuickBird

QuickBird was a high-resolution commercial earth observation satellite, owned by DigitalGlobe launched in 2001 and decayed in 2015. It was the first satellite in a constellation of three scheduled to be in orbit by 2008. QuickBird used Ball Aerospace's Global Imaging System 2000 (BGIS 2000). The satellite collected panchromatic (black and white) imagery at 61 centimeter resolution and multispectral imagery at 2.44- (at 450 km) to 1.63-meter (at 300 km) resolution, as orbit altitude is lowered during the end of mission life.

At this resolution, detail such as buildings and other infrastructure are easily visible. However, this resolution is insufficient for working with smaller objects such as a license plate on a car. The imagery can be imported into remote sensing image processing software, as well as into GIS packages for analysis.

Contractors include Ball Aerospace & Technologies, Kodak and Fokker Space.

QuickBird I

The first QuickBird was launched in November 2000, by EarthWatch from the Plesetsk Cosmodrome in Russia. QB-1 failed to reach planned orbit and was declared a failure.

Prior to QuickBird I and II, DigitalGlobe launched the EarlyBird 1 successfully in 1997 but the satellite lost communications after only four days in orbit due to power system failure.

QuickBird II

QuickBird II (also QuickBird-2 or Quickbird 2), was launched October 18, 2001 from the Vandenberg Air Force Base, California, USA aboard a Boeing Delta II rocket. The satellite was initially expected to collect at 1 meter resolution but after a license was granted in 2000 by the Department of Commerce/NASA, DigitalGlobe was able launch the QuickBird II with 0.61 meter panchromatic and 2.4 meter multispectral (previously planned 4 meter) resolution.

Mission Extension

In April 2011, the Quickbird satellite was raised from an orbit of 450 km to 482 km. The process, started in March 2011, extended the satellite's life. Before the operation the useful life of Quickbird was expected to drop off around mid-2012 but after the successful mission, the new orbit prolonged the satellite life into early 2015.

Decaying

The last picture was acquired on December 17, 2014. On January 27, 2015 QuickBird re-entered Earth's atmosphere.

Specifications

Sensors

- 60 cm (24 in) (1.37 μrad) panchromatic at nadir

- 2.4 m (7 ft 10 in) (5.47 μrad) multispectral at nadir

  o MS Channels: blue (450-520 nm), green (520-600 nm), red (630-690 nm), near-IR (760-890 nm)

Swath Width and Area Size

- Nominal swath width: 18 km at nadir

- Accessible ground swath: 544 km centered on the satellite ground track (to 30° off nadir)

- Area of interest

  o Single area: 18 km by 18 km

  o Strip: 18 km by 360 km

Orbit

- Altitude (original): 450 km – 97.2 degree sun synchronous circular orbit

- Altitude (post-orbit modification): 482 km – 98 degree sun synchronous inclination

- Revisit frequency: 1 to 3.5 days depending on latitude at 60 cm resolution

- Viewing angle: Agile spacecraft, in-track and cross-track pointing

- Period 94.2 minutes

On-board Storage

- 128 Gigabit capacity (approximately 57 single area images)

Spacecraft

- Fueled for 7 years

- 2100 lb (950 kg), 3.04 m (10 ft) in length

Launch

- Launch Date: October 18, 2001

- Launch Window: 1851-1906 GMT (1451-1506 EDT)

- Launch Vehicle: Delta II

- Launch Site: SLC-2W, Vandenberg Air Force Base, California

- USAF Designation: Quickbird 2.

Geo-stationary Satellites

INSAT Program

The Indian National Satellite (INSAT) system is one of the largest domestic communication systems in the Asia-Pacific region. Communication satellites of the INSAT program are placed in Geo-stationary orbits at approximately 36,000 km altitude. The program was established with the commissioning of INSAT-1B in 1983. INSAT space segment consists of 24 satellites out of which 9 are in service (INSAT-3A, INSAT-4B, INSAT-3C, INSAT-3E, KALPANA-1, INSAT-4A, INSAT-4CR, GSAT-8, GSAT-12 and GSAT-10).

GSAT-10

The recent one in the INSAT program, GSAT-10 was launched in September 2012. The satellite orbits in the geo-stationary orbit located at 83^OE longitude. The mission is intended for communication and navigation purposes.

The figure below shows the coverage of the GAGAN payload onboard GSAT-10.

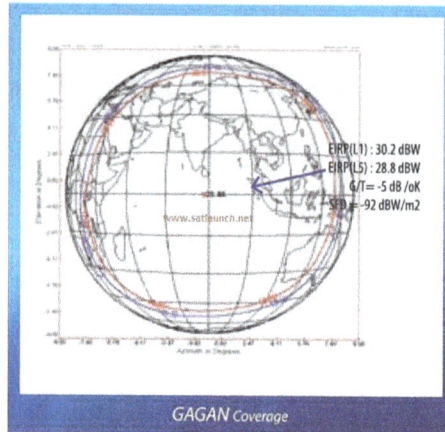

Coverage of the GAGAN payload onboard GSAT-10

Kalpana-1

Another satellite in the INSAT program, KALPANA-1, launched in September 2002, is the first satellite launched by ISRO, exclusively for the meteorological purposes. The satellite orbits in geostationary orbit located at an altitude ~35,786 km and above 74° E longitude. It carries two pay loads: Very High Resolution Radiometer (VHRR) and Data Relay Transponder (DRT). The satellite was originally named Metsat, and was renamed in 2003 in the memory of astronaut Kalpana Chawla.

The VHRR onboard the KALPANA satellite operates in 3 bands: visible, thermal infra-red and water vapour infrared. The instrument gives images in every half an hour.

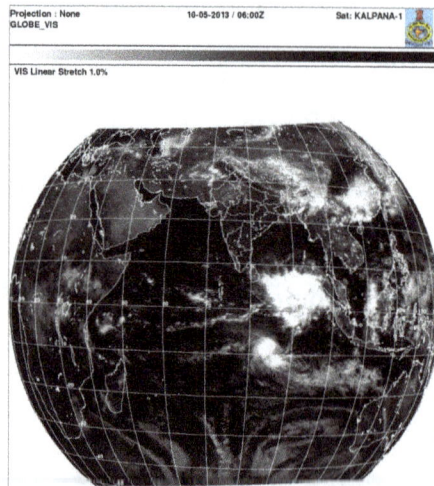

Images from the KALPANA satellite in the Visible spectral band

Images from the KALPANA satellite in the Thermal infrared band

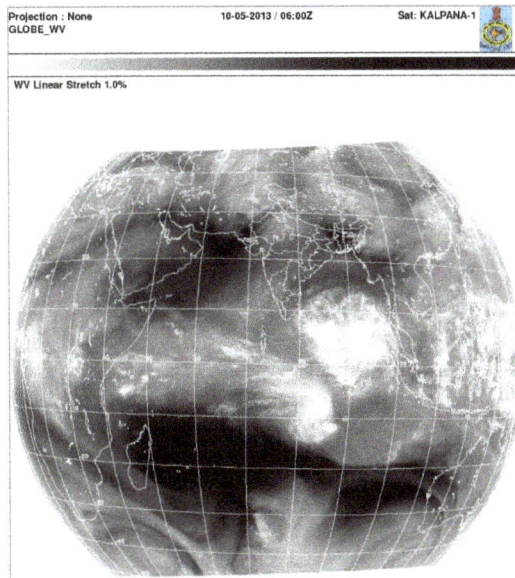

Images from the KALPANA satellite in the Water vapor band

Cartosat

Cartosat series of satellites are examples of earth observation satellites built by India. To date, 4 Cartosat satellites have been built by Indian Space Research Organization (ISRO). Cartosat-1 or IRS-P5 is a stereoscopic earth observation satellite. Maintained by the Indian Space Research Organization (ISRO), this satellite carries two panchromatic (PAN) cameras that take imageries of the earth in the visible region of the electromagnetic spectrum. The imaging capabilities of Cartosat-1 include 2.5 m spatial resolution, 5 day temporal resolution and a 10- bit radiometric resolution.

The second among the series is Cartosat-2 which also images earth using a PAN camera in the visible region of the electromagnetic spectrum. The data obtained has high potential for detailed mapping and other applications at cadastral level. The imaging capabilities of Cartosat-2 are upto 100cm in spatial resolution. The third among the series of satellites was named as Cartosat-2A. This satellite is dedicated for the Indian armed forces. This satellite can be steered upto 45 degrees along as well as across the direction of movement for the purpose of imaging more frequently. Cartosat 2B is the fourth of the Cartosat series of satellites, launched in July 2010. Apart from the imaging capabilities, Cartosat-2B can be steered upto 26 degrees along as well as across the direction of its movement to facilitate frequent imaging of an area. Cartosat-3 is the fifth satellite.

Radarsat

RADARSAT is a constellation of Canadian Remote Sensing satellites that relies on the operational use of Synthetic Aperture Radar (SAR). The main applications for which RADARSAT was designed include:

- Maritime surveillance (such as monitoring of ice, wind, oil pollution etc)

- Disaster management (which includes mitigation, warning, response and recovery)

- Monitoring of ecosystem (such as forestry, agriculture, wetlands etc)

In addition to these applications, RADARSAT offers a wide range of applications involving climate change, land use evolution, coastal change, urban subsidence etc.

References

- "Ariane 5 User's Manual Issue 5 Revision 1" (PDF). arianespace. July 2011. Archived from the original (PDF) on 4 October 2013. Retrieved 28 July 2013

- Michalet, X. (2006). "Using photon statistics to boost microscopy resolution". Proceedings of the National Academy of Sciences. 103 (13): 4797–4798. Bibcode:2006PNAS..103.4797M. PMC 1458746 . PMID 16549771. doi:10.1073/pnas.0600808103

- "Diffraction: Fraunhofer Diffraction at a Circular Aperture" (PDF). Melles Griot Optics Guide. Melles Griot. 2002. Retrieved 2011-07-04

- Lord Rayleigh, F.R.S. (1879). "Investigations in optics, with special reference to the spectroscope". Philosophical Magazine. 5. 8 (49): 261–274. doi:10.1080/14786447908639684

- Turner, Linda (25 October 1995). "Space Imaging granted FCC license for private remote sensing satellite system". Business Wire. Retrieved 23 August 2008

- Schutzberg, Adena (21 September 2005). "ORBIMAGE Acquires Space Imaging: The Past and the Future". Directions Magazine. Retrieved 23 August 2008

- Pohl, D. W.; Denk, W.; Lanz, M. (1984). "Optical stethoscopy: Image recording with resolution $\lambda/20$". Applied Physics Letters. 44 (7): 651. Bibcode:1984ApPhL..44..651P. doi:10.1063/1.94865

- Lockheed Martin (n.d.). "LM900 Bus Program Specification Sheet" (PDF). Archived from the original (PDF) on 2 October 2006. Retrieved 23 August 2008

- "DigitalGlobe Completes Quickbird Satellite Orbit Raise". DigitalGlobe News Room. April 18, 2011. Retrieved 2014-06-11

- "QuickBird 2 was successfully launched on 18 Oct 2001". Center for Remote Imaging, Sensing & Processing. 2001. Retrieved 11 June 2014

Digital Elevation Modelling and its Uses

Digital elevation model (DEM) is the model that is used to represent the surface of a planet in a digital plane. They can be of two types- raster or vector based triangular irregular network. Digital elevation modelling is mostly done with the help of remote sensing and rarely done by using data collected directly. This chapter discusses the methods of digital elevation modelling in a critical manner providing key analysis to the subject matter.

Digital Elevation Modelling

Digital Elevation Model (DEM) is the digital representation of the land surface elevation with respect to any reference datum. DEMs are used to determine terrain attributes such as elevation at any point, slope and aspect. Terrain features like drainage basins and channel networks can also be identified from the DEMs. DEMs are widely used in hydrologic and geologic analyses, hazard monitoring, natural resources exploration, agricultural management etc. Hydrologic applications of the DEM include groundwater modeling, estimation of the volume of proposed reservoirs, determining landslide probability, flood prone area mapping etc.

DEM is generated from the elevation information from several points, which may be regular or irregular over the space. In the initial days, DEMs were used to be developed from the contours mapped in the topographic maps or stereoscopic areal images. With the advancement of technology, today high resolution DEMs for a large part of the globe is available from the radars onboard the space shuttle.

Definition of a DEM

A DEM is defined as "any digital representation of the continuous variation of *relief* over space," (Burrough, 1986), where *relief* refers to the height of earth's surface with respect to the datum considered. It can also be considered as regularly spaced grids of the elevation information, used for the continuous spatial representation of any terrain.

Digital Terrain Model (DTM) and Digital Surface Model (DSM) are often used as synonyms of the DEM. Technically a DEM contains only the elevation information of the surface, free of vegetation, buildings and other non ground objects with reference to a datum such as Mean Sea Level (MSL). The DSM differs from a DEM as it includes the tops of buildings, power lines, trees and all objects as seen in a synoptic view. On the other hand, in a DTM, in addition to the elevation information, several other informations are included, viz., slope, aspect, curvature and skeleton. It thus gives a continuous representation of the smoothed surface.

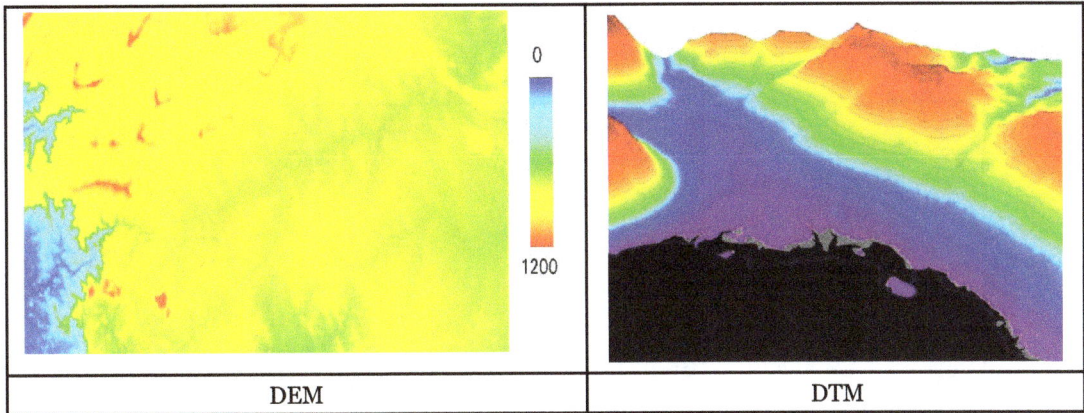

| DEM | DTM |

Example of a (a) DEM and (b) DTM

Types of DEMs

DEMs are generated by using the elevation information from several points spaced at regular or irregular intervals. The elevation information may be obtained from different sources like field survey, topographic contours etc. DEMs use different structures to acquire or store the elevation information from various sources. Three main type of structures used are the following.

a) Regular square grids

b) Triangulated irregular networks (TIN)

c) Contours

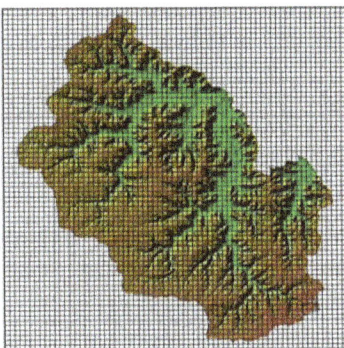

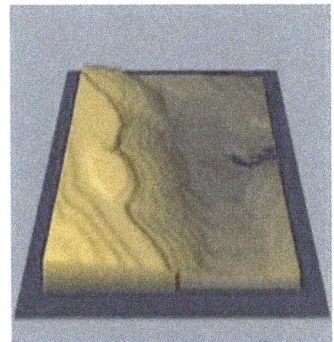

Different types of DEMs (a) Gridded DEM (b) TIN DEM (c) Contour-based DEM

a) Gridded Structure

Gridded DEM (GDEM) consists of regularly placed, uniform grids with the elevation information of each grid. The GDEM thus gives a readily usable dataset that represents the elevation of surface as a function of geographic location at regularly spaced horizontal (square) grids. Since the GDEM data is stored in the form of a simple matrix, values can be accessed easily without having to resort to a graphical index and interpolation procedures.

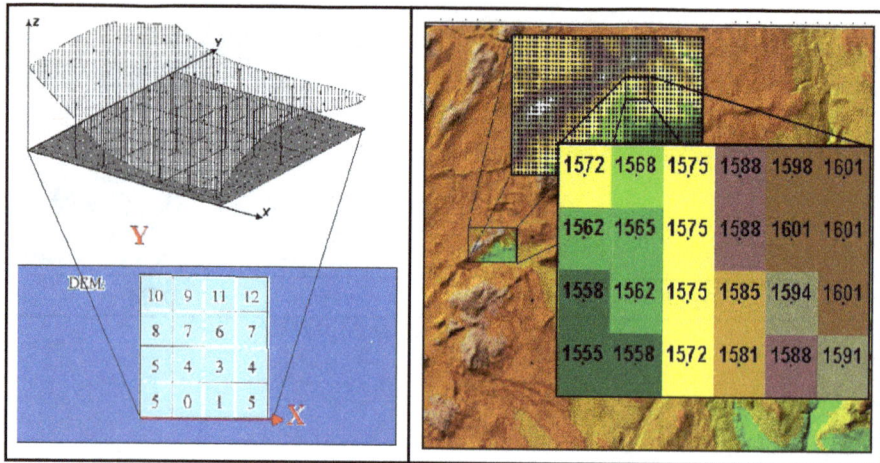

Gridded DEM

Accuracy of the GDEM and the size of the data depend on the grid size. Use of smaller grid size increases the accuracy. However it increases the data size, and hence results in computational difficulties when large area is to be analyzed. On the other hand, use of larger grid size may lead to the omission of many important abrupt changes at sub-grid scale.

Some of the applications of the GDEMs include automatic delineation of drainage networks and catchment areas, development of terrain characteristics, soil moisture estimation and automated extraction of parameters for hydrological or hydraulic modeling.

TIN Structure

TIN is a more robust way of storing the spatially varying information. It uses irregular sampling points connected through non-overlapping triangles. The vertices of the triangles match with the surface elevation of the sampling point and the triangles (facets) represent the planes connecting the points.

Location of the sampling points, and hence irregularity in the triangles are based on the irregularity of the terrain. TIN uses a dense network of triangles in a rough terrain to capture the abrupt changes, and a sparse network in a smooth terrain. The resulting TIN data size is generally much less than the gridded DEM.

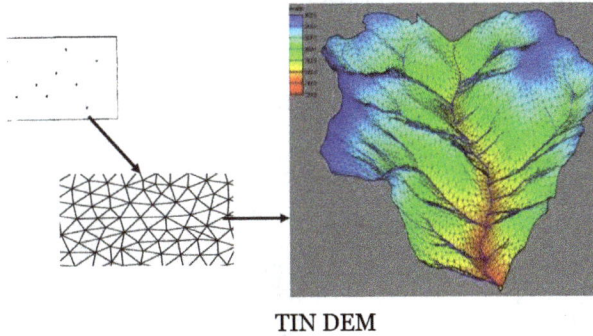

TIN DEM

TIN is created by running an algorithm over a raster to capture the nodes required for the triangles. Even though several methods exist, the Delaunay triangulation method is the most preferred one for generating TIN. TIN for Krishna basin in India created using USGS DEM data is shown in figure. It can be observed from this figure that the topographical variations are depicted with the use of large triangles where change in slope is small. Small triangles of different shapes and sizes are used at locations where the fluctuations in slope are high.

TIN for Krishna basin created from USGS DEM data

Due to its capability to capture the topographic irregularity, without significant increase in the data size, for hydrologic modeling under certain circumstances, TIN DEM has been considered to be better than the GDEM by some researchers (Turcotte et al., 2001). For example, in gridded DEM-based watershed delineation, flow is considered to vary in directions with 45^0 increments. Using TIN, flow paths can be computed along the steepest lines of descent of the TIN facets (Jones et al., 1990).

Contour-based Structure

Contours represent points having equal heights/ elevations with respect to a particular datum such as Mean Sea Level (MSL). In the contour-based structure, the contour lines are traced from the topographic maps and are stored with their location (x, y) and elevation information.

These digital contours are used to generate polygons, and each polygon is tagged with the elevation information from the bounding contour.

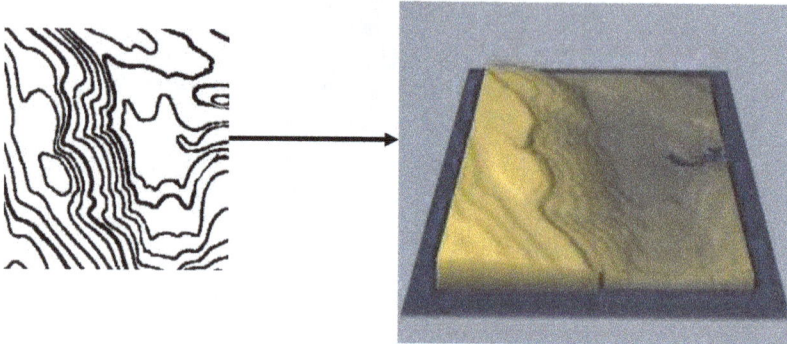

Contour-based DEM

Contour-based DEM is often advantageous over the gridded structure in hydrological and geomorphological analyses as it can easily show the water flow paths. Generally the orthogonals to the contours are the water flow paths.

Major drawback of contour based structure is that the digitized contours give vertices only along the contour. Infinite number of points are available along the contour lines, whereas not many sampling points are available between the contours. Therefore, accuracy of DEM depends on the contour interval. Smaller the contour interval, the better would be the resulting DEM. If the contour interval of the source map is large, the surface model created from it is generally poor, especially along drainages, ridge lines and in rocky topography.

Sources of Digital Elevation Data

Elevation information for a DEM may be acquired through filed surveys, from topographic contours, aerial photographs or satellite imageries using the photogrammetric techniques. Recently radar interferometric techniques and Laser altimetry have also been used to generate a DEM.

Field surveys give the point elevation information at various locations. The points can be selected based on the topographic variations. Contours are the lines joining points of equal elevation. Therefore, contours give elevation at infinite numbers of points, however only along the lines.

A digital elevation model can be generated from the points or contours using various interpolation techniques like linear interpolation, kriging, TIN etc. Accuracy of the resulting DEM depends on the density of data points available depicting the contour interval, and precision of the input data.

On the other hand, photogrammetric techniques provides continuous elevation data using pairs of stereo photographs or imageries taken by instruments onboard an aircraft or space shuttle. Radar interferometry uses a pair of radar images for the same location, from two different points. The difference observed between the two images is used to interpret the height of the location. Lidar altimetry also uses a similar principle to generate the elevation information.

Today very fine resolution DEMs at near global scale are readily available from various sources. The following are some of the sources of global elevation data set.

- GTOPO30

- NOAA GLOBE project

- SRTM

- ASTER Global Digital Elevation Model

- Lidar DEM

GTOPO30 is the global elevation data set published by the United State Geological Survey (USGS). Spatial resolution of the data is 30 arc second (approximately 1 Kilometer).

The Global Land One-km Base Elevation Project (GLOBE) generates a global DEM of 3 arc second (approximately 1 kilometer) spatial resolution. Data from several sources will be combined to generate the DEM. The GLOBE DEM can be obtained from the NOAA National Geophysical Data Centre.

Shuttle Radar Topographic Mission (SRTM) was a mission to generate the topographic data of most of the land surface (56°S to 60°N) of the Earth, which was jointly run by the National Geospatial-Intelligence Agency (NGA) and the National Aeronautics and Space Administration (NASA). In this mission, stereo images were acquired using the Interferometric Synthetic Aperture Radar (IFSAR) instruments onboard the space shuttle Endeavour, and the DEM of the globe was generated using the radar interferometric techniques. The SRTM digital elevation data for the world is available at 3 arc seconds (approximately 90 m) spatial resolution from the website of the CGIAR Consortium for Spatial Information .For the United States and Australia, 30m resolution data is also available.

ASTER Global Digital Elevation Model (GDEM) was generated from the stereo pair images collected by the Advanced Space Borne Thermal Emission and Reflection Radiometer (ASTER) instrument onboard the sun-synchronous Terra satellite. The data was released jointly by the Ministry of Economy, Trade, and Industry (METI) of Japan and the United States National Aeronautics and Space Administration (NASA). ASTER instruments consisted of three separate instruments to operate in the Visible and Near Infrared (VNIR), Shortwave Infrared (SWIR), and the Thermal Infrared (TIR) bands. ASTER GDEMs are generated using the stereo pair images collected using the ASTER instruments, covering 99% of the Earth's land mass (ranging between latitudes 83°N and 83°S). ASTER GDEM is available at 30m spatial resolution in the GeoTIFF format. The ASTER GDEM is being freely distributed by METI (Japan) and NASA (USA) through the Earth Remote Sensing Data Analysis Center (ERSDAC) and the NASA Land Processes Distributed Active Archive Center (LP DAAC) .

Light Detection and Ranging (LIDAR) sensors operate on the same principle as that of laser equipment. Pulses are sent from a laser onboard an aircraft and the scattered pulses are recorded. The time lapse for the returning pulses is used to determine the two-way distance to the object. LIDAR uses a sharp beam with high energy and hence high resolution can be achieved. It also enables DEM generation of a large area within a short period of time with minimum human dependence. The disadvantage of procuring high resolution LIDAR data is the expense involved in data collection.

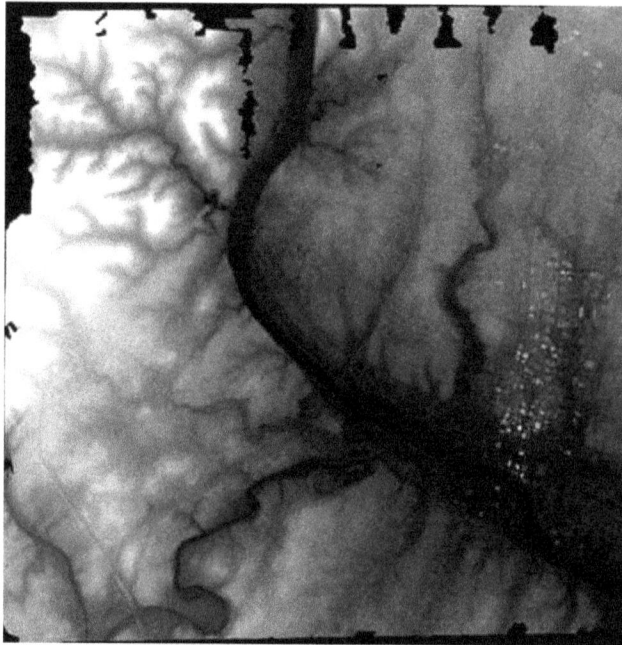

Lidar DEM at 5m resolution for the downtown area of Austin

Radar Interferometry

Radio Detection and Ranging (Radar) is a system which uses microwave signals to detect the presence of an object and its properties. Pulses of radio waves are sent from the radar antenna towards the objects. The objects scatter back a part of this energy falling on them, which are collected at the same radar platform. Energy reflected from the terrain to the radar antenna is called radar return.

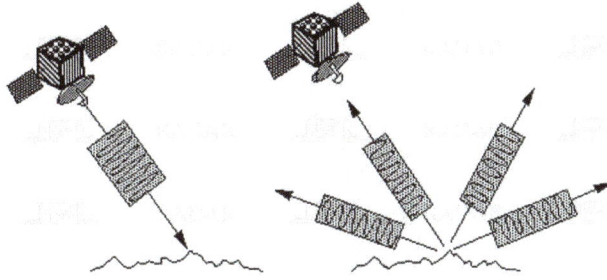

Radar transmits a pulse Measures reflected echo (backscatter)
Principles of radar remote sensing

When an object scatters the radar signal, the scattered signals differ from the original in amplitude, polarization and phase. Thus, the radar information consist complex information in terms of both amplitude and phase of the signals. The difference between the pulses sent and received indicates the properties of the object and the distance of the object.

Principles of radar interferometry are used in the radar remote sensing to generate high resolution DEMs with near global coverage.

Radar Imaging from a Moving Platform

In radar imaging, the radar systems are generally operated from a moving platform, either airborne or space-borne. The radar imaging is based on the Side Looking Airborne Radar (SLAR) systems. In SLAR systems, the microwave pulses are emitted to the side of the aircraft/ space shuttle as it moves forward, and the radar return is recorded using the antenna. Each pulse has a finite duration and it illuminates a narrow strip of land normal to the flight direction as shown in figure. The radar image is generated by using continuous strips swept as the aircraft/ moves in the azimuth direction.

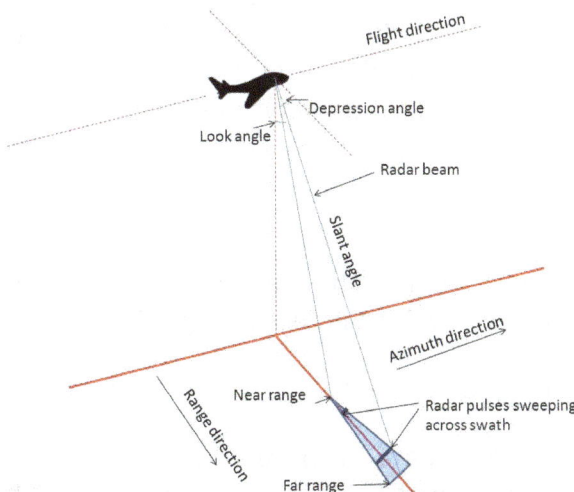

Radar imaging from a moving platform

Characteristics of the Radar Signals

Wavelength

Microwave remote sensing uses the electromagnetic spectrum with wavelength ranging from a few millimeters to 1 m. Each wavelength band is denoted by a letter, as shown in the table below. The bands that are commonly used in the radar remote sensing are highlighted in the Table.

Microwave bands commonly used in remote sensing

Band	Wavelength (cm)
Ka (0.86 cm)	0.8-1.1
K	1.1-1.7
Ku	1.7-2.4
X (3 and 3.2 cm)	2.4-3.8
C	3.8-7.5
S	7.5-15.0
L (25 cm)	15.0-30.0
P	30.0-100.0

(From Sabins, 1978)

The selection of wavelength depends on the objectives of the analysis. Smaller wavelengths cannot penetrate through the clouds and hence are generally less preferred for imaging from airborne/space-borne platforms. Larger wavelengths like L bands are capable of penetrating through the cloud, and hence the satellite-based radar imaging uses the larger wavelength bands.

Longer wavelengths can penetrate through the soil and hence can be used to retrieve soil information. However, they provide less information about the surface characteristics. On the other hand, the shorter wavelengths get scattered from the surface and give more information about the surface characteristics. Hence shorter wavelength bands C and X are used in radar interferometry to extract the topographic information.

Velocity

Microwave bands of the electromagnetic spectrum are used in the radar remote sensing. Therefore these signals travel at the speed of light ($c = 3 \times 10^8$ m.sec^{-1}).

Pulse Duration or Pulse Length

The pulses sent from the radar have a constant duration, which is called the pulse duration or pulse length. The amount of energy transmitted is directly proportional to the pulse length. Resolution of the radar imagery in the range direction is a function of the pulse length. When the pulse length is long, larger area on the ground is scanned by a single pulse, leading to a coarser resolution.

Phase of the Signal

Phase is used to mention the phase of the wave cycle (crest or trough). The phase of the radar return depends on the total distance travelled from the radar to the terrain and back in terms of the total wave cycles.

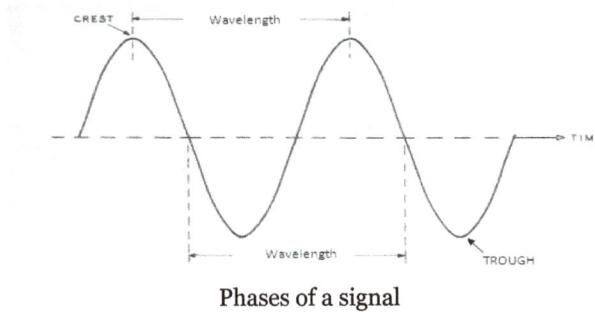

Phases of a signal

Look Angle, Depression Angle and Incident Angle

Look angle is the angle between the nadir and the point of interest on the ground. Depression angle is complementary to the look angle. Angle between the incident radar beam and a line normal to the ground is called the incident angle. These are shown in the figure.

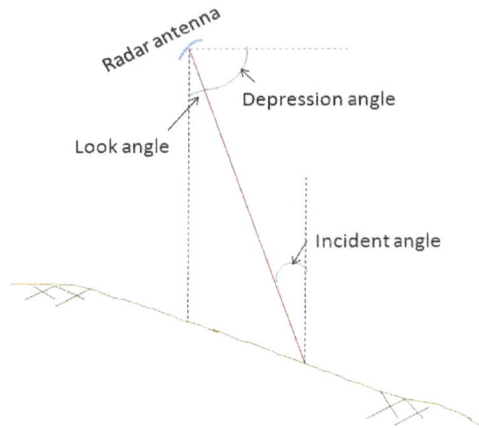

Concept of look angle, incident angle and depression angle in radar remote sensing

Resolution of the Radar Image

Spatial resolution of a radar image is controlled by the pulse length and the antenna beam width. Radar image resolution is specified in terms of both azimuth and range resolutions. A combination of both determines the ground resolution of a pixel.

Range Resolution

In order to differentiate two objects in the range direction, the scattered signals from these two objects must be received at the antenna without any overlap. If the slant distance between two objects is less than half of the pulse length, the reflected signals from the two objects will get mixed and the same will be recorded as the radar return from a single object. The resolution in the range direction is therefore controlled by the pulse duration or pulse length. The slant-range resolution of the radar image is equal to one half of the pulse length (i.e., $\tau/2$).

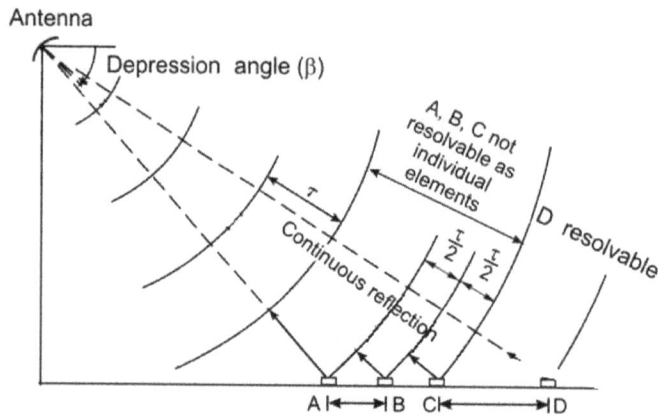

Range resolution of the radar imagery

Ground resolution (Rr) in the range direction, corresponding to the slant range resolution is calculated as follows.

$$R_r = \frac{\tau c}{2\cos(\beta)}$$

where τ is the pulse length measured using the units of time, c is the velocity of the signal (equals velocity of light = 3×10^8 m/sec) and β is the depression angle.

Thus, shorter pulse lengths give better range resolution. But shorter pulses reduce the total energy transmitted, and hence the signal strength. Therefore electronic techniques have been used to shorten the apparent pulse length and hence to improve the resolution.

Azimuth Resolution

Resolution in the azimuth direction (Ra) is controlled by the width of the terrain strip illuminated by the radar beam (or the radar beam width of the antenna), and the ground range. Smaller beam widths give better resolution in the azimuth direction. Since the angular width of the beam is directly proportional to the wavelength λ, and inversely proportional to the length of the antenna (D), resolution is in the azimuth direction is calculated as follows.

$$R_a = \frac{S \cos(\beta) \lambda}{D}$$

where, S is the slant range. Since the radar antenna produces fan shaped beams, in the SLAR the width of the beam is less in the near range and more in the far range as shown in figure. Therefore, the resolution in the azimuth direction is better in the near range, than that in the far range.

Resolution in the azimuth direction improves with the use of shorter wavelengths and larger antennas. Use of shorter waves gives finer resolution in the azimuth direction. However shorter waves carry less energy and hence have poor penetration capacity. Therefore the wave length cannot be reduced beyond certain limits. Also, there are practical limitations to the maximum length of the antenna. Therefore, Synthetic Aperture Radars are used to synthetically increase the antenna length and hence to improve the resolution in the azimuth direction.

Synthetic Aperture Radars

In radar remote sensing, since the spatial resolution is inversely related to the length of the antenna, when the real aperture radars are used, the resolution in the azimuth direction is limited. In Synthetic Aperture Radar (SAR), using the Doppler principle, fine resolution is achieved using both short and long waves.

Doppler Effect

Doppler principal states that if the source or the listener are in relative motion, the frequency of the sound heard differs from the frequency at the source, frequency will be more (or less) depending upon whether the source and the listener are moving close to (or away from) each other, as shown in the figure below.

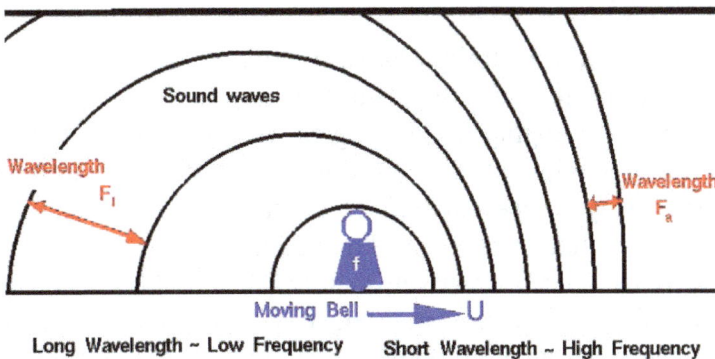

Doppler principle of sound

Application of Doppler principle in SAR

The Doppler principle is used in the SAR to synthesize the effect of a long antenna. In

SAR, each target is scanned using repeated radar beams as the aircraft moves forward, and the radar returns are recorded. In reality, the radar returns are recorded by the same antenna at different locations along the flight track, but these successive positions are treated as if they are parts of a single long antenna, and thereby synthesize the effect of a long antenna.

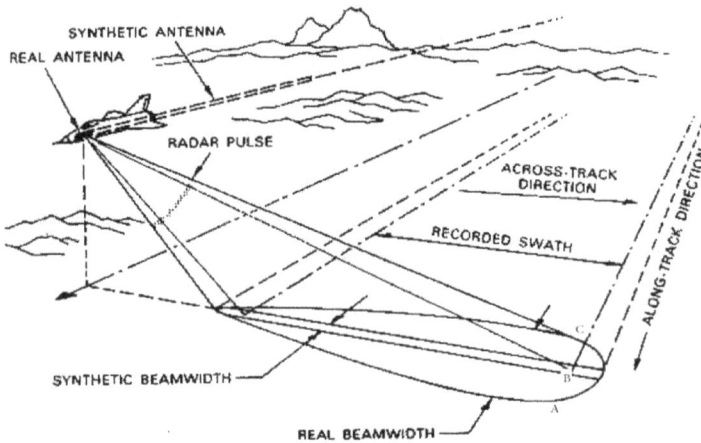

Synthetic Aperture Radar imaging

In SAR, the apparent motion of the target through continuous radar beams is assumed (from A to B and then from B to C) as shown in the figure.When the target is moving closer to the antenna, according to the Doppler principle, the frequency (and hence the energy) of the radar return from the target will be more. On the other hand, when the target is moving away from the antenna, the resulting radar return will be weak. The radar returns are processed according to their Doppler frequency shift, by which very small effective beam width is achieved.

Radar Interferometry

Radar interferometry is the technique used to survey large areas giving moderately accurate values of elevation. The principle of data acquisition in interferometric method is similar to stereo-photographic techniques. When the same area is viewed from different orbits from a satellite at different times, the differences in phase values from the scattered signals may be used to derive terrain information. Radar interferometry makes use of the phase changes of the radar return to measure terrain height. Two or more radar images can be used effectively to generate a DEM. This technique is largely used for hazard monitoring like movement of crustal plates in earthquake prone areas, land subsidence, glacial movement, flood monitoring etc. as they give a high accuracy of upto a centimeter in elevation.

Interference Concept

Interference is the superposition of the waves travelling through the same medium.

Depending upon the phases of the waves superpose, the amplitude of the resultant wave may be higher or lower than the individual waves. When the two waves that meet are in phase i.e., the crest and troughs of the two waves coincide with each other, then the amplitude of the resultant wave will be greater than the amplitude of the individual waves, and this process is called constructive interference. On the other hand, if the two waves are in opposite phase, the amplitude of the resultant wave will be less than that of the individual waves, which is called destructive interference.

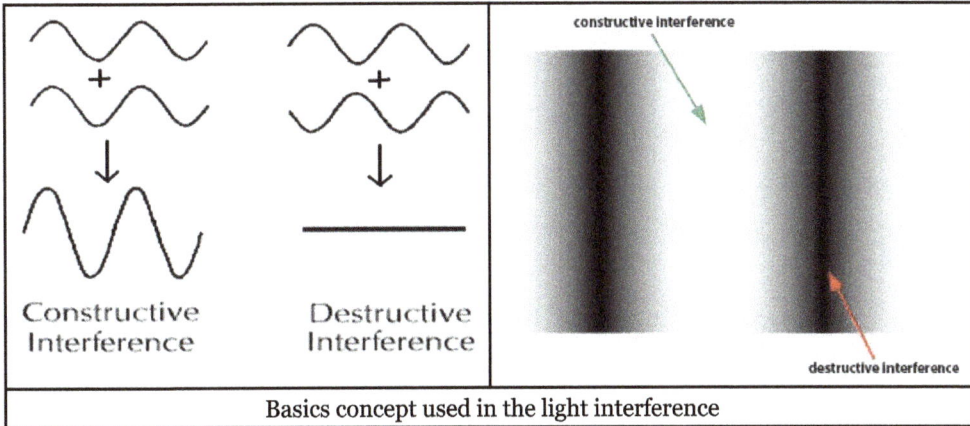

Basics concept used in the light interference

The phase of the resulting signal at any point depends the distance to the point from the source, distance between the sources, and the wavelength. Therefore, if the wavelength of the signal, phase of the resultant signal, and the distance between the two sources are known, the distance of the point from the source can be estimated.

Principles of Radar Interferometry

In radar interferometry, the principles of light interference are used to estimate the terrain height. The principle of data acquisition in interferometric method is similar to stereo- photographic techniques. The radar return from the same object is recorded using two antennas located at two different points in space. Due to the distance between the two antennas, the slant ranges of the radar returns from the same object are different at the two antennas. This difference causes some phase difference between the two radar returns, which may range between 0 and 2π radians. This phase difference information is used to interpret the height of the target.

The phase of the transmitted signal depends on the slant range, and the wavelength of the pulse (λ). Total distance travelled by the pulse is twice the slant range (i.e., $2S$). The phase induced by the propagation of each signal is then given by

$$\theta = 2\pi \frac{2S}{\lambda} = \frac{4\pi S}{\lambda}$$

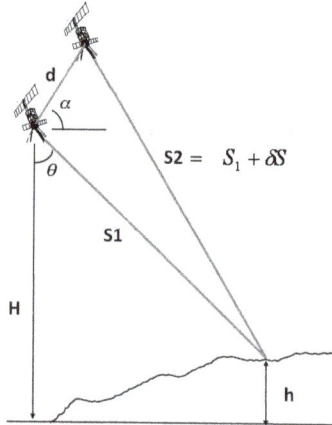

Principle used in the radar interferometry

When the radar returns are recorded at two antennas separated at a distance d as shown in the figure, the difference in the phase is given by

$$\phi_1 - \phi_2 = \frac{4\pi}{\lambda}(S_1 - S_2)$$

When a single antenna is used to send the signals and the radar returns are received at two antennas, the phase difference is given by

$$\phi_1 - \phi_2 = \frac{2\pi}{\lambda}(S_1 - S_2)$$

$$= \frac{2\pi}{\lambda}[(S_1 + S_1) - (S_1 + S_2)]$$

$$= \frac{2\pi}{\lambda}(S_1 - S_2)$$

$$= \frac{2\pi}{\lambda}\delta S$$

In radar interferometry, the images recorded at two antennas are combined to generate the interferogram (also known as fringe map), which gives the interference fringes, as shown in the figure below .

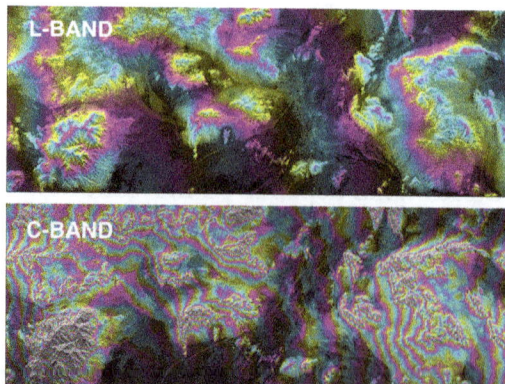

L-band and C-band radar interferogram for the Fort Irwin in California

The interference fringes are used to accurately calculate the phase difference between the two signals for each point in the image. Knowing the wavelength and phase difference, δS can be calculated.

From the principles of trigonometry, the slant range $S1$ can be calculated using the following relation.

$$\operatorname{Sin}(\theta - \alpha) = \frac{(S_1 - \delta S) - S_1^2 - d^2}{2 S_1 d}$$

where d is the base length or the distance between the two antennas, θ is the look angle and α is the angle of inclination of the baseline from the reference horizontal place.

The slant range $S1$ is related to the height of the antenna above the ground, terrain height and the look angle as given below.

$$h = H - S_1 \cos(\theta)$$

Thus, in radar interferometry, knowing the height of the antenna above the ground level (H), look angle (θ), base length (d), inclination of the base from horizontal plane (α), and wavelength of the signal (λ), the measured phase difference is used to estimate the elevation (h) of the terrain.

Types of Radar Interferometry

Single-pass interferometry: Two antennas are located at a known fixed distance apart. Signals are transmitted only from one antenna and the energy scattered back are recorded at both the antennas.

Repeat pass interferometry: In this only one antenna is used to send and receive the signals. The antenna is passed more than once over the area of interest, but through different closely spaced orbits.

Radar Signal Characteristics

A radar system uses a radio frequency electromagnetic signal reflected from a target to determine information about that target. In any radar system, the signal transmitted and received will exhibit many of the characteristics described below.

The Radar Signal in the Time Domain

The diagram below shows the characteristics of the transmitted signal in the time domain. Note that in this and in all the diagrams within this article, the x axis is exaggerated to make the explanation clearer.

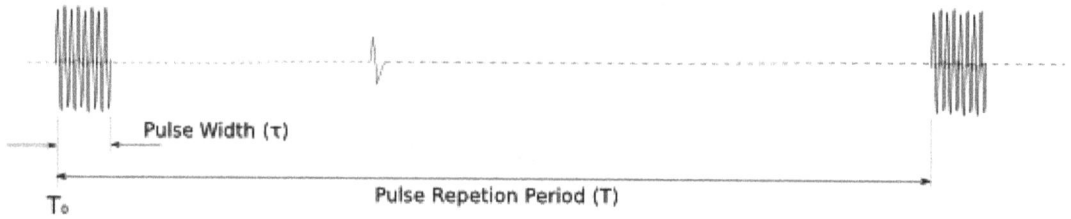

Carrier

The carrier is an RF signal, typically of microwave frequencies, which is usually (but not always) modulated to allow the system to capture the required data. In simple ranging radars, the carrier will be pulse modulated and in continuous wave systems, such as Doppler radar, modulation may not be required. Most systems use pulse modulation, with or without other supplementary modulating signals. Note that with pulse modulation, the carrier is simply switched on and off in sync with the pulses; the modulating waveform does not actually exist in the transmitted signal and the envelope of the pulse waveform is extracted from the demodulated carrier in the receiver. Although obvious when described, this point is often missed when pulse transmissions are first studied, leading to misunderstandings about the nature of the signal.

Pulse Width

The pulse width (τ) (or pulse duration) of the transmitted signal is to ensure that the radar emits sufficient energy to allow that the reflected pulse is detectable by its receiver. The amount of energy that can be delivered to a distant target is the product of two things; the output power of the transmitter, and the duration of the transmission. Therefore, pulse width constrains the maximum detection range of a target.

It also determines the range discrimination, that is the capacity of the radar to distinguish between two targets fairly close together. At *any* range, with similar azimuth and elevation angles and as viewed by a radar with an unmodulated pulse, the range discrimination is approximately equal in distance to half of the pulse duration.

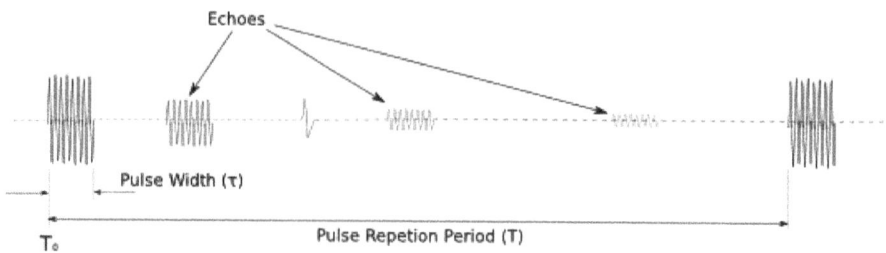

Pulse width also determines the dead zone at close ranges. While the radar transmitter is active, the receiver input is blanked to avoid the amplifiers being swamped (saturated) or, (more likely), damaged. A simple calculation reveals that a radar echo will take approximately 10.8 μs to return from a target 1 statute mile away (counting from the leading edge of the transmitter pulse (T_o), (sometimes known as transmitter main bang)). For convenience, these figures may also be expressed as 1 nautical mile in 12.4 μs or 1 kilometre in 6.7 μs. (For simplicity, all further discussion will use metric figures.) If the radar pulse width is 1 μs, then there can be no detection of targets closer than about 150 m, because the receiver is blanked.

All this means that the designer cannot simply increase the pulse width to get greater range without having an impact on other performance factors. As with everything else in a radar system, compromises have to be made to a radar system's design to provide the optimal performance for its role.

Pulse Repetition Frequency (PRF)

In order to build up a discernible echo, most radar systems emit pulses continuously and the repetition rate of these pulses is determined by the role of the system. An echo from a target will therefore be 'painted' on the display or integrated within the signal processor every time a new pulse is transmitted, reinforcing the return and making detection easier. The higher the PRF that is used, then the more the target is painted. However, with the higher PRF the range that the radar can "see" is reduced. Radar designers try to use the highest PRF possible commensurate with the other factors that constrain it, as described below.

There are two other facets related to PRF that the designer must weigh very carefully; the beamwidth characteristics of the antenna, and the required periodicity with which the radar must sweep the field of view. A radar with a 1° horizontal beamwidth that sweeps the entire 360° horizon every 2 seconds with a PRF of 1080 Hz will radiate 6 pulses over each 1-degree arc. If the receiver needs at least 12 reflected pulses of similar amplitudes to achieve an acceptable probability of detection, then there are three choices for the designer: double the PRF, halve the sweep speed, or double the beamwidth. In reality, all three choices are used, to varying extents; radar design is all about compromises between conflicting pressures.

Staggered PRF

Staggered PRF is a transmission process where the time between interrogations from radar changes slightly, *in a patterned and readily-discernible repeating manner*. The change of repetition frequency allows the radar, on a pulse-to-pulse basis, to differentiate between returns from its own transmissions and returns from other radar systems with the same PRF and a similar radio frequency. Consider a radar with a constant interval between pulses; target reflections appear at a relatively constant range related to the flight-time of the pulse. In today's very crowded radio spectrum, there may be many other pulses detected by the receiver, either directly from the transmitter or as reflections from elsewhere. Because their apparent "distance" is defined by measuring their time relative to the last pulse transmitted by "our" radar, these "jamming" pulses could appear at any apparent distance. When the PRF of the "jamming" radar is very similar to "our" radar, those apparent distances may be very slow-changing, just like real targets. By using stagger, a radar designer can force the "jamming" to jump around erratically in apparent range, inhibiting integration and reducing or even suppressing its impact on true target detection.

Without staggered PRF, any pulses originating from another radar on the same radio frequency might appear stable in time and could be mistaken for reflections from the radar's own transmission. With staggered PRF the radar's own targets appear stable in range in relation to the transmit pulse, whilst the 'jamming' echoes may move around in apparent range (uncorrelated), causing them to be rejected by the receiver. Staggered PRF is only one of several similar techniques used for this, including jittered PRF (where the pulse timing is varied in a less-predictable manner), pulse-frequency modulation, and several other similar techniques whose principal purpose is to reduce the probability of unintentional synchronicity. These techniques are in widespread use in marine safety and navigation radars, by far the most numerous radars on planet Earth today.

Clutter

Clutter refers to radio frequency (RF) echoes returned from targets which are uninteresting to the radar operators. Such targets include natural objects such as ground, sea, precipitation (such as rain, snow or hail), sand storms, animals (especially birds), atmospheric turbulence, and other atmospheric effects, such as ionosphere reflections, meteor trails, and three body scatter spike. Clutter may also be returned from man-made objects such as buildings and, intentionally, by radar countermeasures such as chaff.

Some clutter may also be caused by a long radar waveguide between the radar transceiver and the antenna. In a typical plan position indicator (PPI) radar with a rotating antenna, this will usually be seen as a "sun" or "sunburst" in the centre of the display as the receiver responds to echoes from dust particles and misguided RF in the waveguide. Adjusting the timing between when the transmitter sends a pulse and when the receiver stage is enabled will generally reduce the sunburst without affecting the accuracy of the range, since most sunburst is caused by a diffused transmit pulse reflected before it leaves the antenna. Clutter is considered a passive interference source, since it only appears in response to radar signals sent by the radar.

Clutter is detected and neutralized in several ways. Clutter tends to appear static between radar scans; on subsequent scan echoes, desirable targets will appear to move, and all stationary echoes can be eliminated. Sea clutter can be reduced by using horizontal polarization, while rain is reduced with circular polarization (note that meteorological radars wish for the opposite effect, and therefore use linear polarization to detect precipitation). Other methods attempt to increase the signal-to-clutter ratio.

Clutter moves with the wind or is stationary. Two common strategies to improve measure or performance in a clutter environment are:

- Moving target indication, which integrates successive pulses and

- Doppler processing, which uses filters to separate clutter from desirable signals.

The most effective clutter reduction technique is pulse-Doppler radar with Look-down/shoot-down capability. Doppler separates clutter from aircraft and spacecraft using a frequency spectrum, so individual signals can be separated from multiple reflectors located in the same volume using velocity differences. This requires a coherent transmitter. Another technique uses a moving target indication that subtracts the receive signal from two successive pulses using phase to reduce signals from slow moving objects. This can be adapted for systems that lack a coherent transmitter, such as time-domain pulse-amplitude radar.

Constant False Alarm Rate, a form of Automatic Gain Control (AGC), is a method that relies on clutter returns far outnumbering echoes from targets of interest. The receiver's gain is automatically adjusted to maintain a constant level of overall visible clutter. While this does not help detect targets masked by stronger surrounding clutter, it does help to distinguish strong target sources. In the past, radar AGC was electronically controlled and affected the gain of the entire radar receiver. As radars evolved, AGC became computer-software controlled and affected the gain with greater granularity in specific detection cells.

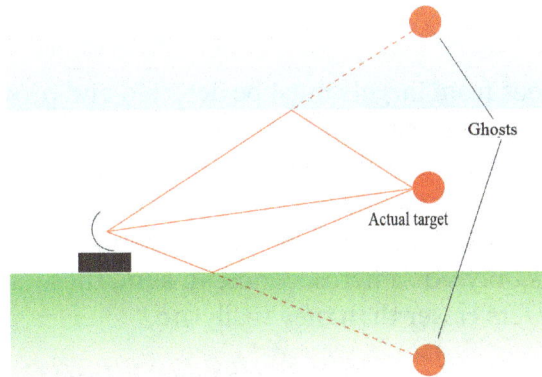

Radar multipath echoes from a target cause ghosts to appear.

Clutter may also originate from multipath echoes from valid targets caused by ground reflection, atmospheric ducting or ionospheric reflection/refraction (e.g., Anomalous propagation). This clutter type is especially bothersome since it appears to move and behave like other normal (point) targets of interest. In a typical scenario, an aircraft echo is reflected from the ground below, appearing to the receiver as an identical target below the correct one. The radar may try to unify the targets, reporting the target at an incorrect height, or eliminating it on the basis of jitter or a physical impossibility. Terrain bounce jamming exploits this response by amplifying the radar signal and directing it downward. These problems can be overcome by incorporating a ground map of the radar's surroundings and eliminating all echoes which appear to originate below ground or above a certain height. Monopulse can be improved by altering the elevation algorithm used at low elevation. In newer air traffic control radar equipment, algorithms are used to identify the false targets by comparing the current pulse returns to those adjacent, as well as calculating return improbabilities.

Sensitivity time Control (STC)

STC is used to avoid saturation of the receiver from close in ground clutter by adjusting the attenuation of the receiver as a function of distance. More attenuation is applied to returns close in and is reduced as the range increases.

Unambiguous Range

Single PRF

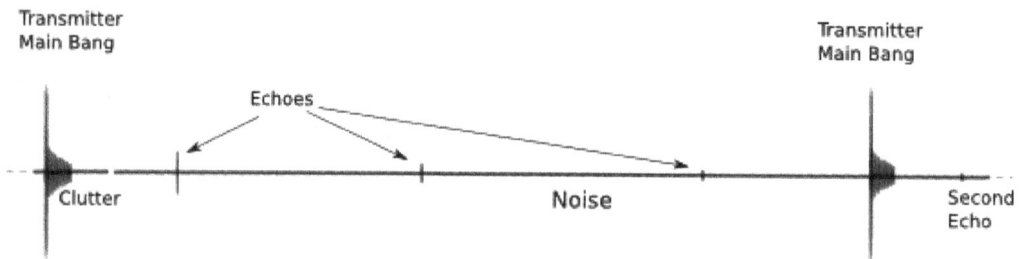

In simple systems, echoes from targets must be detected and processed before the next transmitter pulse is generated if range ambiguity is to be avoided. Range ambiguity occurs when the time taken for an echo to return from a target is greater than the pulse repetition period (T); if the interval between transmitted pulses is 1000 microseconds, and the return-time of a pulse from a distant target is 1200 microseconds, the apparent distance of the target is only 200 microseconds. In sum, these 'second echoes' appear on the display to be targets closer than they really are.

Consider the following example : if the radar antenna is located at around 15 m above sea level, then the distance to the horizon is pretty close, (perhaps 15 km). Ground targets further than this range cannot be detected, so the PRF can be quite high; a radar with a PRF of 7.5 kHz will return ambiguous echoes from targets at about 20 km, or over the horizon. If however, the PRF was doubled to 15 kHz, then the ambiguous range is reduced to 10 km and targets beyond this range would only appear on the display after the transmitter has emitted another pulse. A target at 12 km would appear to be 2 km away, although the strength of the echo might be much lower than that from a genuine target at 2 km.

The maximum non ambiguous range varies inversely with PRF and is given by:

$$\text{Range}_{\text{max unambiguous}} = \left(\frac{c}{2\,PRF} \right)$$

If a longer unambiguous range is required with this simple system, then lower PRFs are required and it was quite common for early search radars to have PRFs as low as a few hundred Hz, giving an unambiguous range out to well in excess of 150 km. However, lower PRFs introduce other problems, including poorer target painting and velocity ambiguity in Pulse-Doppler systems.

Multiple PRF

Modern radars, especially air-to-air combat radars in military aircraft, may use PRFs in the tens-to-hundreds of kilohertz and stagger the interval between pulses to allow the correct range to be determined. With this form of staggered PRF, a *packet* of pulses is transmitted with a fixed interval between each pulse, and then another *packet* is transmitted with a slightly different interval. Target reflections appear at different ranges for each *packet*; these differences are accumulated and then simple arithmetical techniques may be applied to determine true range. Such radars may use repetitive patterns of *packets*, or more adaptable *packets* that respond to apparent target behaviors. Regardless, radars that employ the technique are universally coherent, with a very stable radio frequency, and the pulse *packets* may also be used to make measurements of the Doppler shift (a velocity-dependent modification of the apparent radio frequency), especially when the PRFs are in the hundreds-of-kilohertz range. Radars exploiting Doppler effects in this manner typically determine relative velocity first, from the Doppler effect, and then use other techniques to derive target distance.

Maximum Unambiguous Range

At its most simplistic, MUR (Maximum Unambiguous Range) for a Pulse Stagger sequence may be calculated using the TSP (Total Sequence Period). TSP is defined as the total time it takes for the Pulsed pattern to repeat. This can be found by the addition of all the elements in the stagger sequence. The formula is derived from the speed of light and the length of the sequence:

$$MUR = \left(c * 0.5 * TSP \right)$$

where c is the speed of light, usually in metres per microsecond, and TSP is the addition of all the positions of the stagger sequence, usually in microseconds. However, it should be noted that in a stagger sequence, some intervals may be repeated several times; when this occurs, it is more appropriate to consider TSP as the addition of all the unique intervals in the sequence.

Also, it is worth remembering that there may be vast differences between the MUR and the maximum range (the range beyond which reflections will probably be too weak to be detected), and that the maximum *instrumented* range may be *much* shorter than either of these. A civil marine radar, for instance, may have user-selectable maximum *instrumented* display ranges of 72, or 96 or rarely 120 nautical miles, in accordance with international law, but maximum unambiguous ranges of over 40,000 nautical miles and maximum detection ranges of perhaps 150 nautical miles. When such huge disparities are noted, it reveals that the primary purpose of staggered PRF is to reduce "jamming", rather than to increase unambiguous range capabilities.

The Radar Signal in the Frequency Domain

Pure CW radars appear as a single line on a Spectrum analyser display and when modulated with other sinusoidal signals, the spectrum differs little from that obtained with standard analogue modulation schemes used in communications systems, such as Frequency Modulation and consist of the carrier plus a relatively small number of sidebands. When the radar signal is modulated with a pulse train as shown above, the spectrum becomes much more complicated and far more difficult to visualise.

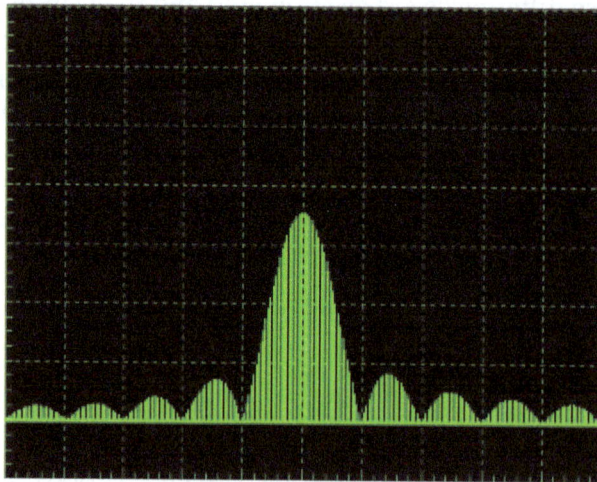

Basic Fourier analysis shows that any repetitive complex signal consists of a number of harmonically related sine waves. The radar pulse train is a form of square wave, the pure form of which consists of the fundamental plus all of the odd harmonics. The exact composition of the pulse train will depend on the pulse width and PRF, but mathematical analysis can be used to calculate all of the frequencies in the spectrum. When the pulse train is used to modulate a radar carrier, the typical spectrum shown on the left will be obtained.

Examination of this spectral response shows that it contains two basic structures. The Coarse Structure; (the peaks or 'lobes' in the diagram) and the Fine Structure which contains the individual frequency components as shown below. The Envelope of the lobes in the Coarse Structure is given by: $\frac{1}{\pi f}$.

Note that the pulse width (τ) determines the lobe spacing. Smaller pulse widths result in wider lobes and therefore greater bandwidth.

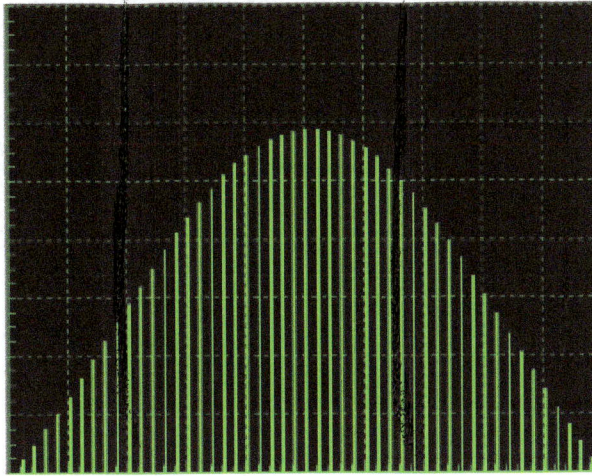

Examination of the spectral response in finer detail, as shown on the right, shows that the Fine Structure contains individual lines or spot frequencies. The formula for the fine structure is given by $\frac{N}{T}$ and since the period of the PRF (T) appears at the bottom of the fine spectrum equation, there will be fewer lines if higher PRFs are used. These facts affect the decisions made by radar designers when considering the trade-offs that need to be made when trying to overcome the ambiguities that affect radar signals.

Pulse Profiling

If the rise and fall times of the modulation pulses are zero, (e.g. the pulse edges are infinitely sharp), then the sidebands will be as shown in the spectral diagrams above. The bandwidth consumed by this transmission can be huge and the total power transmitted is distributed over many hundreds of spectral lines. This is a potential source of interference with any other device and frequency-dependent imperfections in the transmit chain mean that some of this power never arrives at the antenna. In reality of course, it is impossible to achieve such sharp edges, so in practical systems the sidebands contain far fewer lines than a perfect system. If the bandwidth can be limited to include relatively few sidebands, by rolling off the pulse edges intentionally, an efficient system can be realised with the minimum of potential for interference with nearby equipment. However, the trade-off of this is that slow edges make range resolution poor. Early radars limited the bandwidth through filtration in the transmit chain, e.g. the waveguide, scanner etc., but performance could be sporadic with unwanted signals breaking through at remote frequencies and the edges of the recovered pulse being indeterminate. Further examination of the basic Radar Spectrum shown above shows that the information in the various lobes of the Coarse Spectrum is identical to that contained in the main lobe, so limiting the transmit and receive bandwidth to that extent provides significant benefits in terms of efficiency and noise reduction.

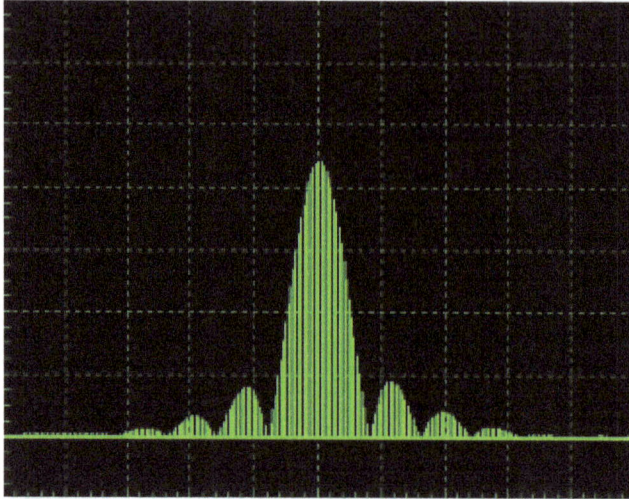

Recent advances in signal processing techniques have made the use of pulse profiling or shaping more common. By shaping the pulse envelope before it is applied to the transmitting device, say to a cosine law or a trapezoid, the bandwidth can be limited at source, with less reliance on filtering. When this technique is combined with pulse compression, then a good compromise between efficiency, performance and range resolution can be realised. The diagram shows the effect on the spectrum if a trapezoid pulse profile is adopted. It can be seen that the energy in the sidebands is significantly reduced compared to the main lobe and the amplitude of the main lobe is increased.

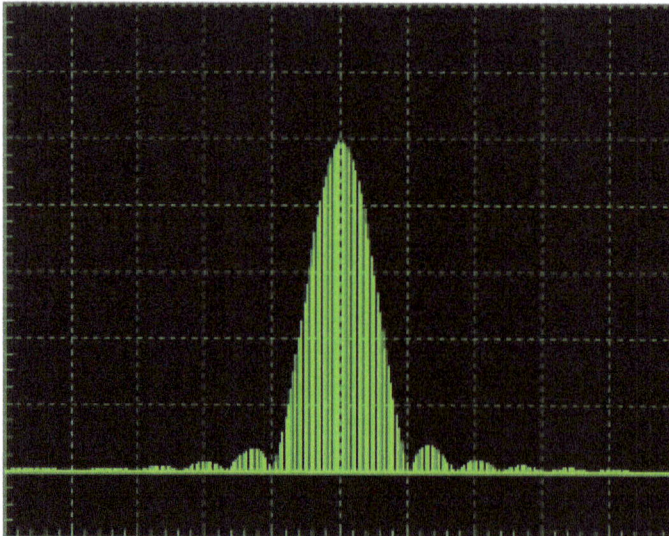

Similarly, the use of a cosine pulse profile has an even more marked effect, with the amplitude of the sidelobes practically becoming negligible. The main lobe is again increased in amplitude and the sidelobes correspondingly reduced, giving a significant improvement in performance.

There are many other profiles that can be adopted to optimise the performance of the system, but cosine and trapezoid profiles generally provide a good compromise between efficiency and resolution and so tend to be used most frequently.

Unambiguous Velocity

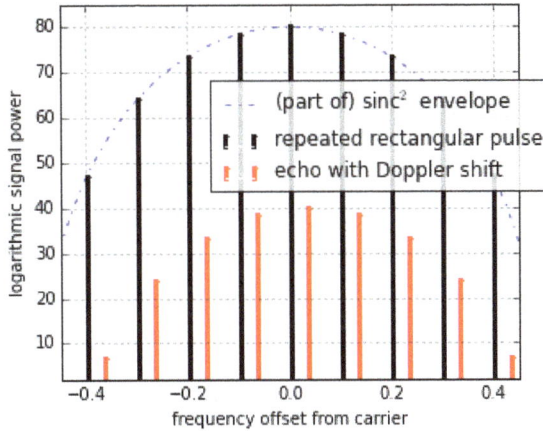

This is an issue only with a particular type of system; the Pulse-Doppler radar, which uses the Doppler effect to resolve velocity from the apparent change in frequency caused by targets that have net radial velocities compared to the radar device. Examination of the spectrum generated by a pulsed transmitter, shown above, reveals that each of the sidebands, (both coarse and fine), will be subject to the Doppler effect, another good reason to limit bandwidth and spectral complexity by pulse profiling.

Consider the positive shift caused by the closing target in the diagram which has been highly simplified for clarity. It can be seen that as the relative velocity increases, a point will be reached where the spectral lines that constitute the echoes are hidden or aliased by the next sideband of the modulated carrier. Transmission of multiple pulse-packets with different PRF-values, e.g. staggered PRFs, will resolve this ambiguity, since each new PRF value will result in a new sideband position, revealing the velocity to the receiver. The maximum unambiguous target velocity is given by:

$$\pm \frac{c\,PRF}{4f}$$

Typical System Parameters

Taking all of the above characteristics into account means that certain constraints are placed on the radar designer. For example, a system with a 3 GHz carrier frequency and a pulse width of 1 μs will have a carrier period of approximately 333 ps. Each transmitted pulse will contain about 3000 carrier cycles and the velocity and range ambiguity values for such a system would be:

PRF	Velocity Ambiguity	Range Ambiguity
Low (2 kHz)	50 m/s	75 km
Medium (12 kHz)	300 m/s	12.5 km
High (200 kHz)	5000 m/s	750 m

Shuttle Radar

Shuttle Radar Topographic Mission Data

Availability of a reasonably accurate elevation information for many parts of the world was once very much limited. Dense forest, high mountain ranges etc. were remained unmapped, mainly because of the difficulty in getting to these places. Objective of the Shuttle Radar Topographic Mission (SRTM) was to create near global data set on land elevations, using radar images. The mission was headed by the National Geospatial-Intelligence Agency (NGA) and the National Aeronautics and Space Administration (NASA).

The space shuttle Endeavour was employed in the mission to carry the payloads to the space. The space shuttle Endeavour with the SRTM payloads was launched on 11th February 2000. The Endeavour orbited the Earth at an altitude of 233 km, with 59 deg. inclination, and the radar onboard the space shuttle was used to collect the images of the land surface. The mission was completed in 11 days. These radar images were interpreted to generate a high resolution elevation data, at a near global scale.

Radar system is advantageous over the optical systems as it can operate day and night, and in bad weather. Also, by using space-borne radar system for the mapping, the accessibility issues are eliminated. Thus in the SRTM, around 80% of the land areas were swiped using the radar and the digital elevation data was generated.

The near-global elevation data generated by the SRTM finds extensive applications in the areas of earth system sciences, hydrologic analyses, land use planning, communication system designing, and military purposes.

Instruments Onboard the Payload of the SRTM

The SRTM instruments consisted of two antennas. One antenna was located at the bay of the space shuttle. A mast of 60 m length was connected to the main antenna truss and the second antenna was connected at the end of the mast as shown in figure. The mast provided the baseline distance between the two antennas.

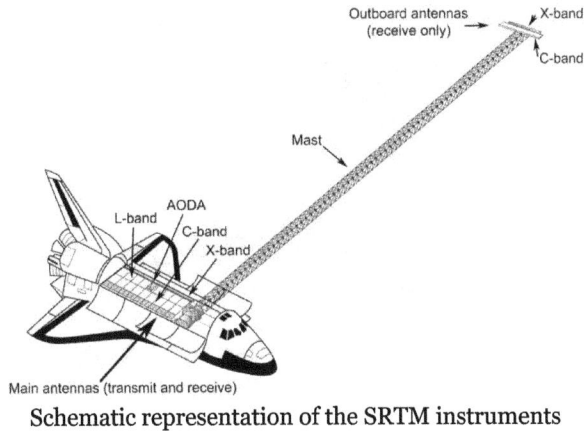

Schematic representation of the SRTM instruments

Main Antenna

The main antenna consisted of two antennas to work in two different wavelengths. The two microwave bands used in the SRTM were the C band and X band.

C-band Radar Antenna

The C-band antenna could transmit and receive radar signals of wavelength 5.6 centimeters. The swath width of the (width of the radar beam on Earth's surface) C-band antenna was 225 kilometers. C-band data was used to scan about 80% of the land surface of the Earth (between $60^{\circ}N$ and $56^{\circ}S$) to produce near-global topographic map of the Earth.

X-band Radar Antenna

The X-band antenna was used to transmit and receive radar signals of wavelength 3 centimeters. Using the shorter wavelengths, X-band radar could achieve higher resolution compared to C-band radar. However, the swath width of the X-band radar was only 50 km. Therefore, the X-band radar could not achieve near global coverage during the mission.

The Mast

The mast was used to maintain the baseline distance between the main antenna and the outboard antenna. The length of the mast was 60 m and was inclined at 45 deg. from the vertical.

Outboard Antenna

The outboard antenna was connected to the end of the mast. It was used only to receive the radar signals scattered back from the land surface. No signal was transmitted from the outboard antenna.

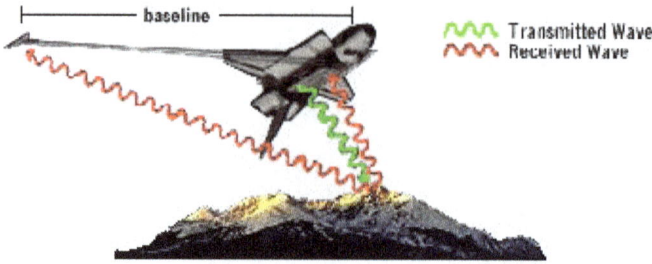

Radar signals being transmitted and recieved in the SRTM mission
(image not to scale).

Radar signals transmitted and received in the SRTM mission

The outboard antenna also contained two antennas: one was used to receive radar signals in the C-band, and the other in the X-band. Wavelengths of the C and X band signals were 5.6 cm and 3 cm, respectively.

How SRTM Generates the Elevation Data

In SRTM, principle of radar interferometry was used to extract the elevation data.

In SRTM, space-borne, fixed baseline, single-pass interferometry was adopted, in which the signal was sent from a single source and the energy scattered back (radar return) was recorded simultaneously using two antennas placed at a fixed known distance apart.

The main antenna located at the bay of the space shuttle was used to send the radar signals. The radar return was recorded at both the main antenna and the outboard antenna (located at 60m away from the main antenna using the mast).

The following figure shows the schematic representation of the SRTM radar system employed for capturing the topographic information.

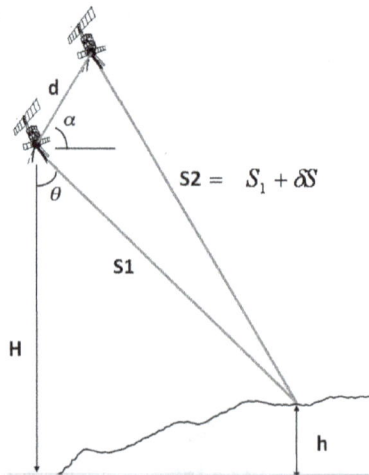

Schematic representation of the STRM imaging

The images recorded at the two antennas were combined to generate the interferogram (or the interference fringes). Since the two antennas were separated by a fixed distance, the radar returns from an object recorded at these two antennas differed in phase, depending upon the distance of the target from the radar antenna. The inteferogram was used to accurately calculate the phase difference between two signals for each point in the image.

Knowing the wavelength of the signal (which were 5.6 cm and 3cm for the C and X bands, respectively), fixed base length (which was 60m) and the phase difference, the slant range S of the object was calculated, by using the principles of trigonometry. Further, the elevation (h) of the target was calculated as shown below.

$$h = H - S_1 \cos(\theta)$$

Where, H is the height of the antenna from the ground level, which in this case is the altitude of the orbit (233 km). The parameter θ is the look angle of the radar signal.

Processes Involved in the SRTM

Various steps involved in the generation of SRTM elevation data are shown in the figure below.

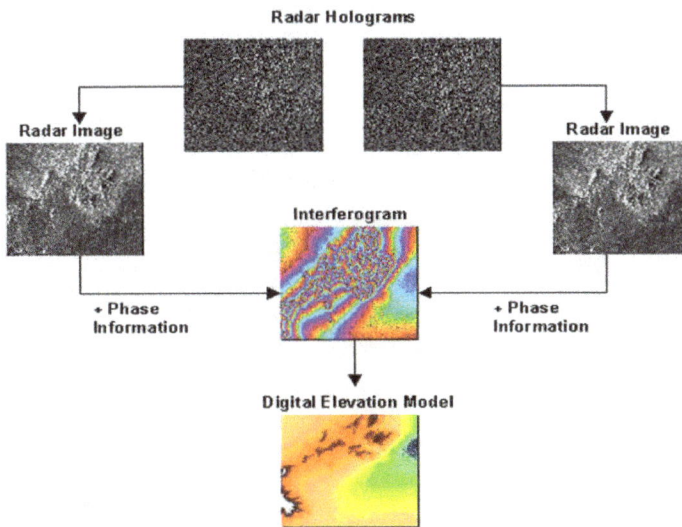

Steps involved in the SRTM elevation data generation

Two radar antennas were used to simultaneously capture the radar returns and the radar images or the radar holograms were created. The radar holograms recorded at the main antenna and the outboard antenna were combined to generate the interferogram (or the fringe map), which displays bands of colors depending up on the phase difference (interferometric phase) between the signals received at the two antennas. The phase difference was then used to calculate the difference in the distance of the target from the two antennas, which was further used to estimate the height of the target.

SRTM Elevation Data Details

SRTM digital elevation data was generated from radar signals using the principles of radar interferometry. This elevation data was then edited to fill small voids, remove spikes, delineate and flatten water bodies, and to improve the coastlines.

The C-band antennas were used to scan almost 80% of the land surface of the Earth (between 60°N and 56°S) to produce the near-global topographic map of the Earth at a spatial resolution of 1 arc-seconds. Due to the smaller swath with of the X-band antennas, near global coverage could not be achieved using the X-band.

The following figures show the coverage of C-band and X-band elevation data, respectively

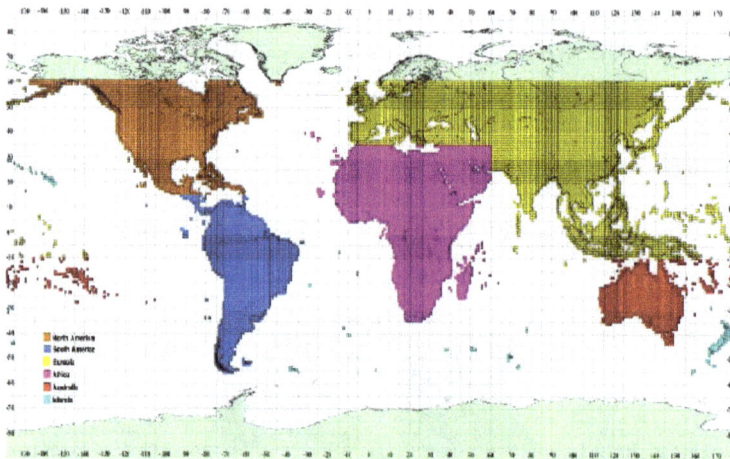

Coverage of the SRTM C-band elevation data

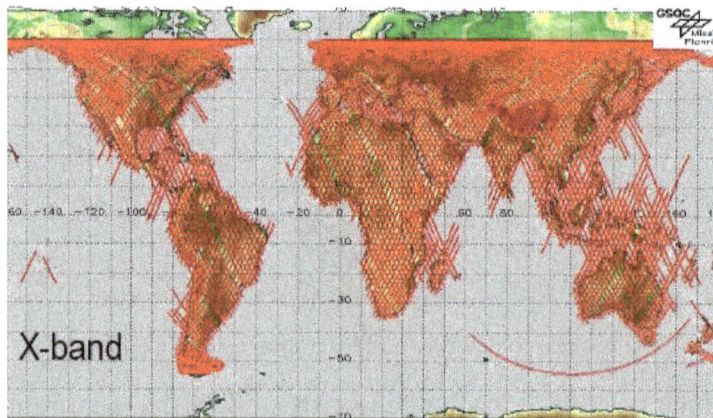

Coverage of the SRTM X-band elevation data

SRTM provides processed DEM at two different spatial resolutions, 1 arc-second (approximately 30 m) only for the Unites States and Australia, and reduced 3 arc-second (approximately 90 m) for 80% of the globe. The data can be obtained from the USGS Earth Explorer website.

The SRTM elevation data is available in two formats.

- Digital Terrain Elevation Data (DTED): Each file contains regularly spaced grids containing the elevation information. This format can be directly used in the GIS platform.

- Band interleaved by line (BIL): In this format, the elevation data is available for regular grids, in binary format. A header file is used to describe the layout of the grids and the data format.

The publically available SRTM DEM is geo-referenced using WGS84 datum. In the data set, unit of elevation is meters. The elevation data gives less than 16m accuracy in the absolute elevation and less than 10m accuracy in the relative vertical height, at 90% confidence level.

Practical Use of SRTM Data in the Tropics

A total of 5 case studies are being provided that analyse the quality, accuracy and usability of SRTM data. It was shown that SRTM DEM is by far a great improvement on the previously existing global DEM products like GTOPO30. Studies were also conducted to analyse the precision of SRTM data in the tropics using GPS data. SRTM DEM was found to be more accurate with errors systematically related to aspect. Data from GPS were compared with SRTM elevation information for a single catchment. The study concludes that at the catchment scale, high quality 1:10,000 cartographic maps produce a more detailed DEM than the 92-m SRTM DEM even though the vertical errors were similar. Case study results are provided which look at the problem of missing data holes in SRTM DEM. The errors introduced due to an interpolation technique for filling SRTM holes were shown. A detailed hydrological analysis of the case study region showed that overall, the interpolation technique for filling missing data holes perform quite well in representing the hydrological characteristics of the catchment.

Case study results present by Jarvis conclude that SRTM derived DEMs provide greater accuracy than TOPO DEMs. At the same time, this does not necessarily mean that it contains more details. It was suggested that 3-arc second SRTM DEMs failed to capture topographic features at the scales of 1:25,000 and below. Hence, presence of cartography with scales above 1:25,000 (eg., 1:50,000 and 1:100,000) implied usage of SRTM DEMs. For hydrological modeling application, if good quality cartography data of the sale 1:25,000 and below are available, digitizing and interpolating this cartographic data was deemed suitable for better results. This is because even though SRTM 3-arc second DEMs perform well for hydrological applications, these are on the margin of usability.

Primary Attributes and Secondary Attributes

Estimation of Attributes from Raster Dem

Terrain attributes derived from the DEM are broadly classified as primary attributes and secondary attributes. Primary attributes are those derived directly from the DEM, whereas secondary attributes are derived using one or more of the primary attributes. Some of the primary attributes, which are important in the hydrologic analysis, derived from the DEM include slope, aspect, flow-path length, and upslope contributing area. Topographic wetness index is an example of the secondary attribute derived from the DEM. Topographic wetness index represents the extent of the zone of saturation as a function of the upslope contributing area, soil transmissivity and slope.

Gridded DEM represents the surface as a matrix of regularly spaced grids carrying the elevation information. Most of the terrain analysis algorithms using the gridded DEM assume uniform spacing of grids throughout the DEM. Topographic attributes are derived based on the changes in the surface elevation with respect to the distance.

Calculation of Slope from the DEM

Slope is defined as the rate of change of elevation, expressed as gradient (in percentage) or in degrees.

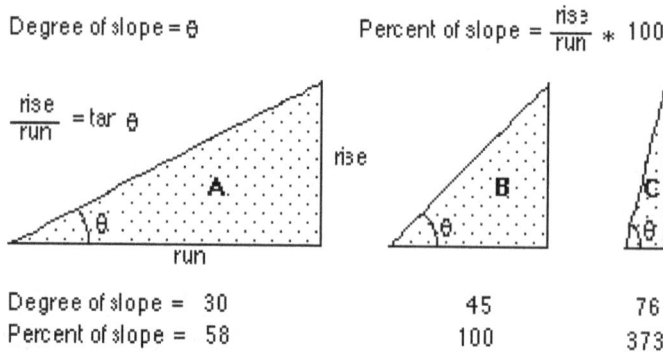

Degree of slope $= \theta$ Percent of slope $= \dfrac{rise}{run} * 100$

$\dfrac{rise}{run} = tan\ \theta$

Degree of slope =	30	45	76
Percent of slope =	58	100	373

Estimation of slope

Using the finite difference approach, slope in any direction is expressed as the first derivative of the elevation in that direction.

Slope in the x direction $S_x = \dfrac{\partial h_x}{\partial x}$

Slope in the y direction $S_y = \dfrac{\partial h_y}{\partial x}$

In general, slope of any point is given as follows

$$S = \sqrt{\left(\frac{\partial h_x}{\partial x}\right)^2 + \left(\frac{\partial h_y}{\partial y}\right)^2}$$

Consider the example grids given below, in which the elevation are marked as h1, h2... h9.

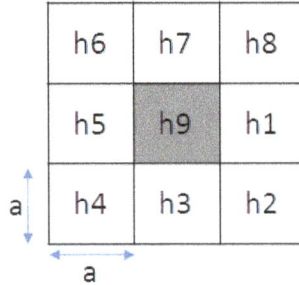

h6	h7	h8
h5	h9	h1
h4	h3	h2

Using the finite difference formulation, slope of the central grid (grid 9) can be calculated as follows.

$$S_9 = \sqrt{\left(\frac{h1 - h5}{2a}\right)^2 + \left(\frac{h3 - h7}{2a}\right)^2}$$

In this approach, only the four directions are considered. However, the slope estimated using the above equation generally gives reasonably accurate values of the slope.

On the other hand, deterministic eight-neighborhood (D8) algorithm estimates the slope by calculating the rate of change of elevation in the steepest down slope direction among the 8 nearest neighbors, as shown below.

$$S_{9,D8a \; lg \; otithm} = Max\left[\frac{h9 - h1}{L}, \frac{h9 - h2}{L}, ..., \frac{h9 - h8}{L}\right]$$

where L is equal to a for cardinal neighbors and $\sqrt{2}\,a$ for diagonal neighbors. This approach is generally preferred when the channel slope is required.

When gridded DEMs are used for the analysis, each of the grids is compared with its nearest 8 neighbors and the slope is derived for each grid. The local slope calculated from a gridded DEM decreases with increase in the DEM grid size. This is because, as the grid size increases, the grids represent larger areas. In the slope calculation, since the spatial averages of the elevation for such larger areas are used, it tends to result in smoother or less steep surface (Wolock and Price 1994).

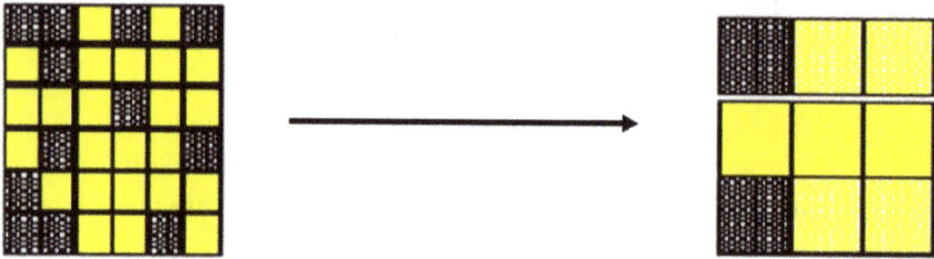

DEM grid size and slope relationship

Selection of DEM grid size is therefore important to get appropriate slope map, particularly in fields like erosion studies, where the processes are largely related to the slope.

Calculation of aspect from the DEM

Aspect is the orientation of the line of steepest descent, normally measured clockwise from the north, and is expressed in degrees.

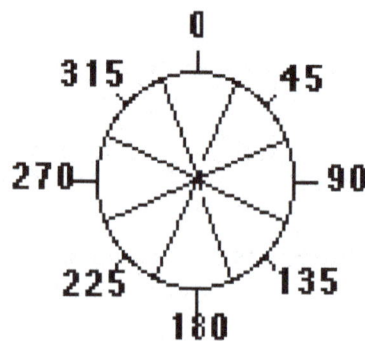

Aspect directions

Determination of Flow Direction

In watershed analysis using raster based DEM, water from each cell is assumed to flow or drain into one of its eight neighboring cells which are towards left, right, up, down, and the four diagonal directions. The flow vector algorithm scans each cell of the DEM, and determines the direction of the steepest downward slope to an adjacent cell.

The most common method used for identifying the flow direction is the D8 (deterministic eight-neighbors) method. The method was first proposed by O'Callaghan and Mark (1984). In this method, a flow vector indicating the steepest slope is assigned to one of the eight neighboring cells.

A detailed description of the D8 algorithm for drainage pattern extraction is provided in the following sub-section.

D8 Algorithm

In this method, the flow direction for each cell is estimated from elevation differences between the given cell and its eight neighboring cells, and hence the name D8 algorithm. Most GIS implementations use the D8 algorithm to determine flow path.

In D-8 algorithm, water from each cell is assumed to flow or drain into one of its eight neighboring cells which are towards left, right, up, down, and the four diagonal directions.

The flow is assumed to follow the direction towards the cell having the steepest slope. If the steepest downward slope is encountered at more than one adjacent cell, flow direction is assigned arbitrarily towards the first of these cells encountered in a row by row scan of the adjacent cells. Where an adjacent cell is undefined (i.e. has a missing elevation value or lies outside the DEM grid), the downward slope to that cell is assumed to be steeper than that to any other adjacent cell with a defined elevation value.

Once the flow direction is identified, numerical values are assigned to the direction. The general flow direction code or the eight-direction pour point model, followed for each direction from the center cell is shown in the figure below. Each of the 8 flow directions are assigned numeric values, using the 2^X series, where $x = 0, 1, 2$ ....etc.

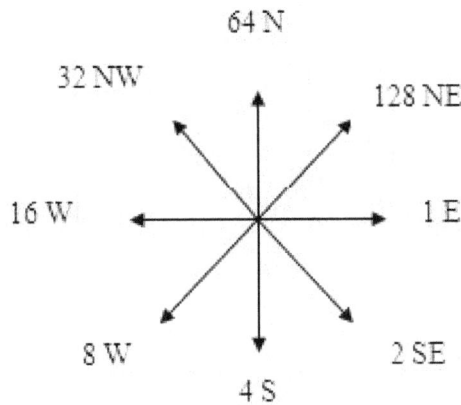

The eight direction pour point model

From the figure, it can be inferred that the flow direction is coded as 1 if the direction of steepest drop is to the right of the current processing cell, and 128 if the steepest drop is towards top- right of the current cell.

Consider a small part of a gridded DEM as shown in the figure. Assume the cell size as 1 unit in dimension. Consider the grid with elevation 67. There are 3 adjacent grids having elevation less than 67 (with elevation 56, 53 and 44), and these three are considered as the possible flow directions. In the flow vector algorithm, the direction of steepest slope among these three directions is identified.

Slope is calculated along the three directions as shown in the figure (b). Slope is the maximum towards the bottom right cell, and hence water would follow that direction. As a result, according to the eight direction pour point model, the center cell (with elevation 67) is allotted a flow direction value of 2.

The same procedure is repeated for all the grids in a DEM. The resulting flow directions and the corresponding flow direction values are shown in the figure (c) and (d), respectively.

(a)

78	72	69	71	58
74	67	56	49	46
69	53	44	37	38
64	58	55	22	31
68	61	47	21	16

(b)

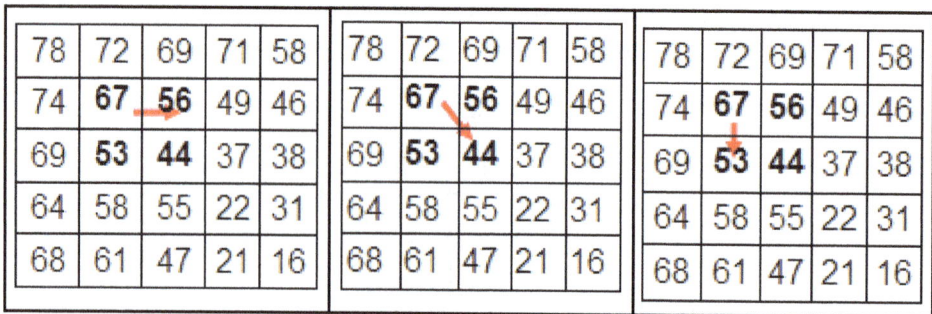

$$Slope = \frac{67 - 56}{1} = 11.0 \qquad Slope = \frac{67 - 44}{\sqrt{2}} = 16.26 \qquad Slope = \frac{67 - 53}{1} = 14.00$$

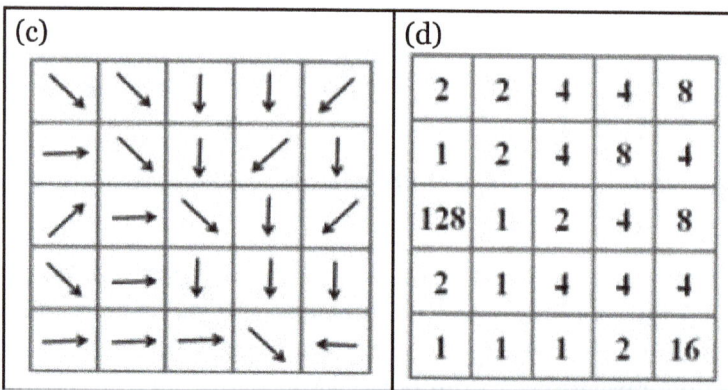

(a) A sample DEM (b) Estimation of the steepest down slope direction (c) Flow direction (d) Flow direction matrix with numerical values for each direction

Disadvantages of eight direction pour point method are the following:

- It limits the direction of flow between two adjacent nodes to only eight possibilities

- There is discrepancy between the lengths of the drainages as calculated by the method

- The method fails to capture parallel flow lines

Use of two outflow paths (Tarboton, 1997), partitioning of the flow into all down slope (Quinn et al., 1991), use of a stochastic approach to determine the gradient (Fairfield and Leymarie, 1991) were some of the improvements made on the D-8 algorithms latter.

Drainage Pattern and Catchment Area Delineation

Topography of the river basin plays an important role in hydrologic modelling, by providing information on different terrain attributes which enhance the assessment and enable the simulation of complex hydrological processes. In the past, topographical maps were one of the major sources of information for derivation of the catchment characteristics in hydrological models. With the rapidly increasing availability of topographic information in digital form, like digital elevation models (DEMs), the automatic extraction of terrain attributes to represent the catchment characteristics has become very popular. Automatic algorithms used for the extraction of the catchment characteristics are benefitted by the speed, accuracy and standardization. The ability to create new, more meaningful characteristics, thereby eliminating the need to draw and digitize the attributes, is another major advantage of such algorithms.

Hydrologic models use information about the topography in the form of terrain attributes that are extracted from DEM for modeling the different hydrological process such as interception, infiltration, evaporation, runoff, groundwater recharge, water quality etc. In hydrologic studies DEMs have also been used to derive the channel network and to delineate the catchment area.

Drainage Pattern Extraction from DEM

Gridded DEM has been widely used in the hydrologic modeling to extract drainage patterns of a basin required for flow routing in the hydrologic models. The gridded DEM provides elevation information at regularly spaced grids over the area. The algorithm used must be capable of identifying the slope variation and possible direction of flow of water using the DEM.

While using the gridded DEM, inadequate elevation difference between the grids often creates difficulty in tracing the drainage pattern. Also, gridded DEM may contain depressions, which are grids surrounded by higher elevations in all directions. Such depressions may be natural or sometimes interpolation errors. These depressions also create problems in tracing the continuous flow path.

Prior to the application of the DEM in the hydrologic studies, preprocessing of the DEM is therefore carried out to correct for the depressions and flat areas.

Treatment of Depressions and Flat Area

Depression or sink is defined here as a point which is lower than its eight nearest neighboring grids, as shown in the figure. Such points may arise due to data errors introduced in the surface generation process, or they represent real topographic features such as quarries or natural potholes.

42	42	42	42	44
41	41	41	42	44
40	37	41	42	44
39	39	40	42	44
38	38	39	41	43

Example for a spurious depression in a gridded DEM

Spurious flat areas are present in a DEM when the elevation information is inadequate to represent the actual relief of the area. The following shows example of spurious flat area in a raster DEM.

43	43	43	43	43
43	43	43	42	43
43	43	43	42	43
42	42	42	42	42
42	41	41	41	42

Example for a spurious flat area in a gridded DEM

There are many algorithms available in literature for treating depressions and flat areas in the raster DEM.

The depression filling algorithms basically identify and delineate the depression. Outlet from the depression is identified by filling the depression to the lowest value on its rim (Jenson and Domingue, 1988).

Advanced algorithms available for the depression filling make use of the overall drainage information of the watershed, either by burning the stream network (overlay the stream network and modify the elevation along channel grids) or by tracing the flow direction from the watershed outlet to the depression.

A DEM which is free of sinks is termed as a depressionless DEM.

Relief algorithms are used to identify the flow direction through flat areas in the DEM. The relief algorithm imposes relief over the flat areas to allow an unambiguous definition of flow lines across these areas.

Two implicit assumptions in the relief algorithm are the following:

- The flat areas are not truly level, but have a relief that is not detectable at the vertical resolution of the original DEM.

- The relief in the flat area is such that any flow entering or originating from that area will follow the shortest path over the flat area to a point on its perimeter where a downward slope is available.

The relief algorithm proposed by Martz and Garbrecht (1998), and the priority-first-search algorithm (PFS) proposed by Jones (2002), are some of the methods available for treating the flat areas and to trace the flow paths.

A DEM, made free of sinks and flat areas is termed as a modified DEM.

Care should be taken as these algorithms could change the natural terrain, enlarge the depression, loop the depression and/or produce an outflow point in the depression while processing a flat area which in turn could affect watershed delineation, drainage network extraction and hydrologic event simulation accuracy.

The steps for delineating watershed from a depressionless DEM are the following.

i. Identification of the flow direction for each grid

ii. Delineation of the flow network

iii. Calculation of flow accumulation at each grid

iv. Stream network delineation

v. Delineation of the stream links

Determination of Flow Vectors

Flow vector algorithms scan each cell of the modified DEM (from the depressions and flat areas) and determine the direction of the steepest downward slope to an adjacent cell. Most common method used for identifying the flow direction is the D8 (deterministic eight- neighbors) method.

Using D-8 algorithm, flow direction for each cell is estimated from elevation differences between the given cell and its eight neighboring cells.

Consider a small sample of a DEM in raster format given in Fig.(a). The corresponding flow direction grid and the matrix containing numerical values of the flow directions are shown in figure (b) and (c), respectively.

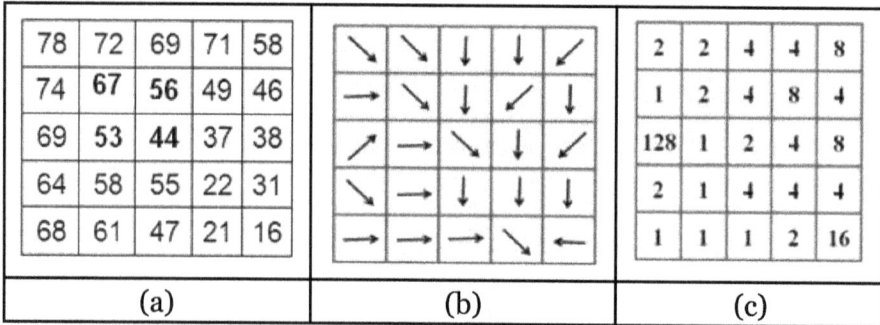

(a) A sample DEM (b) Flow direction grid (d) Flow direction matrix with numerical values for each direction

Flow Network

Once flow direction grid has been obtained, flow network is created by extending the lines of steepest descent beyond each cell as shown in figure

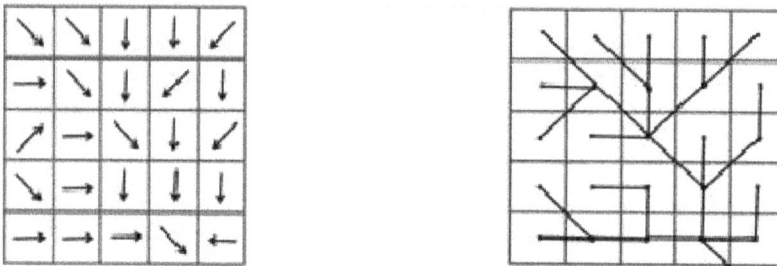

(a) Flow directions and (b) Flow network

Flow Accumulation Grid

Once the flow directions are identified, flow from each grid is traced to the watershed outlet. During this flow tracing, for each grid, a counter is initiated for each grid. As the flow passes through each grid, this counter is incremented by 1. Using the counter, total number of upstream grids that flow into each grid are identified and the flow accumulation grid is generated. In the flow accumulation grid, the number in each cell denotes the number of the cells that flow into that particular cell. Figure (b) clearly illustrates the flow accumulation grid.

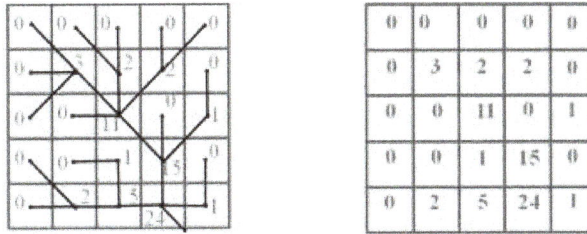

Figure(a) Flow network and (b) Flow accumulation grid

Delineation of the Stream Network

Stream network is defined using the relative count of grids flowing to it, which is obtained from the flow accumulation matrix. To delineate a stream from the flow accumulation grid, it is necessary to specify a threshold flow accumulation. The threshold specifies the minimum flow accumulation at which a grid can be considered as a part of the stream. Grids which have flow accumulation greater than the threshold value are assumed to be the parts of the stream and the remaining grids are considered as overland flow grids.

Grids that are parts of the stream network using the threshold
flow accumulation of 5-cells

Delineation of the Stream Links

On specifying the threshold flow accumulation, the stream links associated with this threshold are obtained with the help of flow network grid, as highlighted in the following figure. Knowing the stream links, the grids contributing flow to any point on the stream can be identified, which can be used to delineate the subwatersheds.

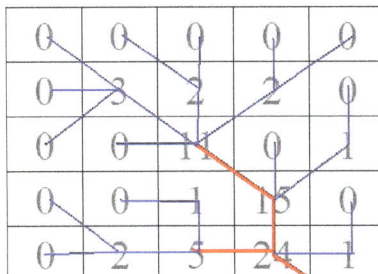

Stream network for a 5-cell threshold flow accumulation (shown in red color)

Watershed Delineation from the DEM

Watershed boundaries or the location of watershed divides can be obtained based on stream channel information. Usually, watershed boundaries tend to be half way between the stream of interest and all the other neighbouring streams. If a much more precise boundary needs to be determined, the use of topographic maps are essential. These maps show elevation contours, small ephemeral streams, large water bodies etc. Pour point is the name given to the outlet of the watershed that user is interested in delineating. An estimate of watershed boundary from a topographic map requires the pour point locations. Pour points can be any point on the stream/river where the surface flow from watershed exists. The upslope catchment area at each grid of the modified gridded DEM is determined using the flow direction information. Beginning at each grid with a defined elevation value and using the flow direction vectors previously generated, the path of steepest descent is continuously followed for each grid until the edge of the DEM is reached, and the flow accumulation is derived for each grid. The flow accumulation represents the number of grids contributing flow into the grid in the watershed, and hence gives the upslope catchment area for that grid.

All the above steps to extract sub-watersheds from a raster based DEM are shown in the form of a flowchart in the following figure.

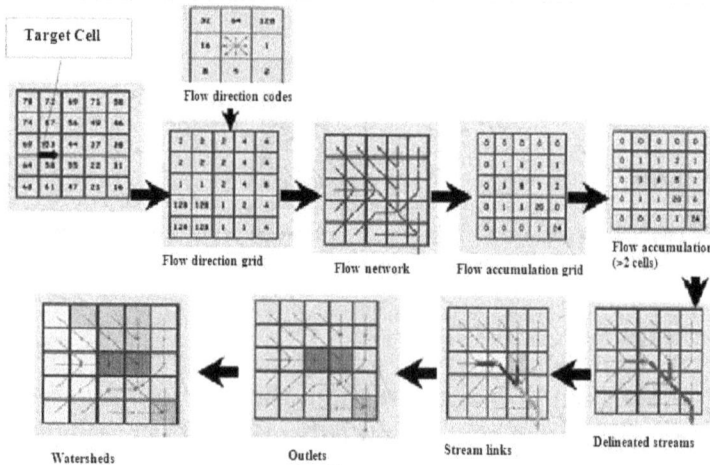

Steps in extracting sub-watersheds from DEM using D8 algorithm

Remote Sensing: Areas of Application

Remote sensing is applied in various fields. Some of these areas are flood mapping, environmental monitoring, irrigation management and watershed. Remote sensing helps in generating awareness related to social and economic problems like deforestation, land usage and discovering natural resources. The topics discussed in the section are of great importance to broaden the existing knowledge on remote sensing.

Watershed

Scientific planning and management is essential for the conservation of land and water resources for optimum productivity. Watersheds being the natural hydrologic units, such studies are generally carried out at watershed scale and are broadly referred under the term watershed management. It involves assessment of current resources status, complex modeling to assess the relationship between various hydrologic components, planning and implementation of land and water conservation measures etc.

Remote sensing via aerial and space-borne platforms acts as a potential tool to supply the essential inputs to the land and water resources analysis at different stages in watershed planning and management. Water resource mapping, land cover classification, estimation of water yield and soil erosion, estimation of physiographic parameters for land prioritization and water harvesting are a few areas where remote sensing techniques have been used.

Various remote sensing applications in water resources management under the following five classes:

- Water resources mapping
- Estimation of watershed physiographic parameters
- Estimation of hydrological and meteorological variables
- Watershed prioritization
- Water conservation

Water Resources Mapping

Identification and mapping of the surface water boundaries has been one of the simplest and direct applications of remote sensing in water resources studies. Water resources mapping using remote sensing data require fine spatial resolution so as to achieve accurate delineation of the boundaries of the water bodies.

Optical remote sensing techniques, with their capability to provide very fine spatial resolution have been widely used for water resources mapping. Water absorbs most of the energy in NIR and MIR wavelengths giving darker tones in the bands, and can be easily differentiated from the land and vegetation.

The following figure shows images of a part of the Krishna river basin in different bands of the Landsat ETM+. In the VIS bands (bands 1, 2 and 3) the contrast between water and other features are not very significant. On the other hand, the IR bands (bands 4 and 5) show a sharp contrast between them due to the poor reflectance of water in the IR region of the EMR spectrum.

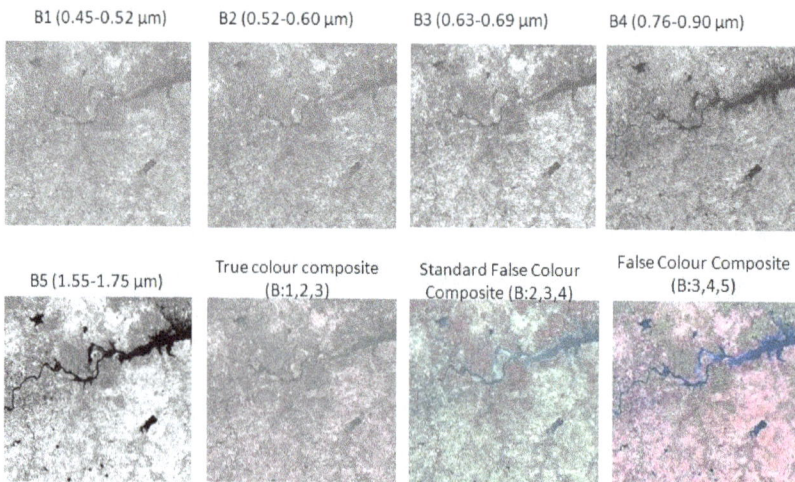

Landsat ETM+ images of a part of the Krishna river basin in different spectral bands
(Nagesh Kumar and Reshmidevi, 2013)

Poor cloud penetration capacity and poor capability to map water resources under thick vegetation cover are the major drawbacks of the optical remote sensing techniques.

Use of active microwave sensor helps to overcome these limitations as the radar waves can penetrate the clouds and the vegetation cover to some extent. In microwave remote sensing, water surface provides specular reflection of the microwave radiation, and hence very little energy is scattered back compared to the other land features. The difference in the energy received back at the radar sensor is used for differentiating, and to mark the boundaries of the water bodies.

Estimation of Watershed Physiographic Parameters

This section covers the remote sensing applications in estimating watershed physiographic parameters and the land use / land cover information.

Watershed Physiographic Parameters

Various watershed physiographic parameters that can be obtained from remotely sensed data include watershed area, size and shape, topography, drainage pattern and landforms.

Stereoscopic attribute of aerial photographs or satellite images permit quantitative assessment of landforms and evaluation of basin topography, which can be used to develop or update the topographic maps. With the help of satellite remote sensing, global scale digital elevation models (DEMs) are available today at fine spatial resolution and reasonable vertical accuracy. DEM from the Shuttle Radar Topographic Mission (SRTM) and ASTER GDEM are examples. SRTM DEM provides near-global DEM at 90m spatial resolution and 16m vertical accuracy. Airborne laser altimeters also provide quick and accurate measurements for evaluating changes in land surface features and are effective tools to ascertain watershed properties.

Fine resolution DEMs have been used to extract the drainage network/ pattern using the flow tracing algorithms. The drainage information can also be extracted from the optical images using digital image processing techniques.

The drainage information may be further used to generate secondary information such as structure of the basin, basin boundary, stream orders, stream length, stream frequency, bifurcation ratio, stream sinuosity, drainage density and linear aspects of channel systems etc.

The figure below shows the ASTER GDEM for a small region in the Krishna Basin in North Karnataka and the drainage network delineated from it using the flow tracing algorithm included in the 'spatial analyst' tool box of ArcGIS. Fig.(b) also shows the stream orders assigned to each of the delineated streams.

(a) ASTER GDEM of a small region in the Krishna Basin (b) and the stream network delineated from the DEM

Land use / Land Cover Classification

Detailed land use / land cover map is another important input that remote sensing can yield for hydrologic analysis.

Land cover classification using multispectral remote sensing data is one of the earliest, and well established remote sensing applications in water resources studies. With the capability of the remote sensing systems to provide frequent temporal sampling and the fine spatial resolution, it is possible to analyze the dynamics of land use / land cover pattern, and also its impact on the hydrologic processes.

Use of hyper-spectral imageries helps to achieve further improvement in the land use / land cover classification, wherein the spectral reflectance values recorded in the narrow contiguous bands are used to differentiate different land use classes which show close resemblance with each other. Identification of crop types using hyper-spectral data is an example.

With the help of satellite remote sensing, land use land cover maps at near global scale are available today for hydrological applications. European Space Agency (ESA) has released a global land cover map of 300 m resolution, with 22 land cover classes at 73% accuracy.

Global 300 m land cover classification from the European Space Agency

Estimation of Hydrological and Meteorological Variables

Hydrological processes such as precipitation and evapotranspiration are generally used as inputs to the hydrological models to simulate other processes such as runoff (surface and sub- surface), storage change in the unsaturated zone, and ground water flow. This section covers the remote sensing applications in estimating precipitation, evapotranspiration and soil moisture.

Precipitation

Remote sensing techniques have been used to provide information about the occurrence of rainfall and its intensity. Basic concept behind the satellite rainfall estimation is the differentiation of precipitating clouds from the non-precipitating clouds (Gibson and Power, 2000) by relating the brightness of the cloud observed in the imagery to the rainfall intensities.

Satellite remote sensing uses both optical and microwave remote sensing (both passive and active) techniques.

The following table lists some of the important satellite rainfall data sets, satellites used for the data collection and the organizations that control the generation and distribution of the data.

Table: Details of some of the important satellite rainfall products (Nagesh Kumar and Reshmidevi, 2013)

Program	Organization	Spectral bands used	Characteristics and source of data
World Weather Watch	WMO	VIS, IR	1-4 km spatial, and 30 min. temporal resolution
TRMM	NASA JAXA	VIS, IR Passive & active microwave	Sub-daily 0.25^{o} (~27 km) spatial resolution
PERSIANN	CHRS	IR	0.25^{o} spatial resolution Temporal resolution: 30 min. aggregated to 6 hrs.
CMORPH	NOAA	Microwave	0.08 deg (8 km) spatial and 30 min. temporal resolution
Acronyms			
CHRS: Center for Hydrometeorology and Remote Sensing, University of California, USA CMORPH: Climate Prediction Center (CPC) MORPHing technique			
NASA: National Aeronautics and Space Administration, USA NOAA: National Oceanic and Atmospheric Administration, USA			
PERSIANN: Precipitation Estimation from Remotely Sensed Information using Artificial Neural Network TRMM: Tropical Rainfall Measuring Mission			
WMO: World Meteorological Organization			

Evapotranspiration

Evapotranspiration (ET) represents the water and energy flux between the land surface and the lower atmosphere. ET fluxes are controlled by the feedback mechanism between the atmosphere and the land surface, soil and vegetation characteristics, and the hydro- meteorological conditions.

There are no direct methods available to estimate the actual ET by means of remote sensing techniques. Remote sensing application in the ET estimation is limited to the estimation of the surface conditions like albedo, soil moisture, surface temperature, and vegetation characteristics like normalized differential vegetation index (NDVI) and leaf area index (LAI). The data obtained from remote sensing are used in different models to simulate the actual ET.

Courault et al. (2005) grouped the remote sensing data-based ET models into four different classes:

- Empirical direct methods: Use the empirical equations to relate the difference in the surface air temperature to the ET.

- Residual methods of the energy budget: Use both empirical and physical parameterization. Example: SEBAL (Bastiaanssen et al., 1998), FAO-56 method (Allen at al., 1998)

- Deterministic models: Simulate the physical process between the soil, vegetation and atmosphere making use of remote sensing data such as Leaf Area Index (LAI) and soil moisture. SVAT (Soil-Vegetation-Atmosphere-Transfer) model is an example (Olioso et al., 1999).

- Vegetation index methods: Use the ground observation of the potential or reference ET. Actual ET is estimated from the reference ET by using the crop coefficients obtained from the remote sensing data (Allen et al., 2005; Neale et al., 2005).

Optical remote sensing using the VIS and NIR bands have been commonly used to estimate the input data required for the ET estimation algorithms.

As a part of the NASA / EOS project to estimate global terrestrial ET from earth's land surface by using satellite remote sensing data, MODIS Global Terrestrial Evapotranspiration Project (MOD16) provides global ET data sets at regular grids of 1 sq.km for the land surface at 8-day, monthly and annual intervals for the period 2000-2010.

Soil Moisture Estimation

Remote sensing techniques of soil moisture estimation are advantageous over the conventional *in-situ* measurement approaches owing to the capability of the sensors to capture spatial variation over a large aerial extent. Moreover, depending upon the revisit time of the satellites, frequent sampling of an area and hence more frequent soil moisture measurements are feasible.

The following figure shows the global average monthly soil moisture in May extracted from the integrated soil moisture database of the European Space Agency- Climate Change Initiative (ESA-CCI).

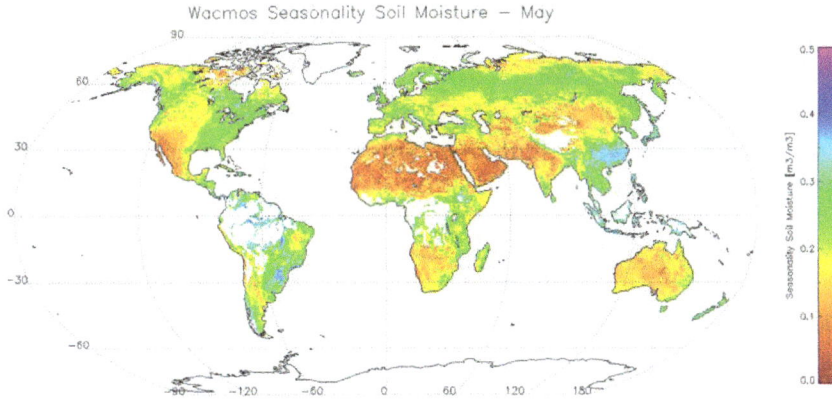

Global monthly average soil moisture in May from the CCI data

Remote sensing of the soil moisture requires information below the ground surface and therefore mostly confined to the use of thermal and microwave bands of the EMR spectrum.

Remote sensing of the soil moisture is based on the variation in the soil properties caused due to the presence of water. Soil properties generally monitored for soil moisture estimation include soil dielectric constant, brightness temperature, and thermal inertia.

Though the remote sensing techniques are giving reasonably good estimation of the soil moisture, due to the poor surface penetration capacity of the microwave signals, it is considered to be effective in retrieving the moisture content of the surface soil layer of maximum 10 cm thickness. In the recent years, attempts have been made to extract the soil moisture of the entire root zone with the help of remote sensing data. Such methods assimilate the remote sensing derived surface soil moisture data with physically based distributed models to simulate the root zone soil moisture. For example, Das et al. (2008) used the Soil-Water-Atmosphere-Plant (SWAP) model for simulating the root zone soil moisture by assimilating the aircraft-based remotely sensed soil moisture into the model.

Some of the satellite based sensors that have been used for retrieving the soil moisture information are the following.

- Passive microwave sensors: SMMR, AMSR-E and SSM/I

- Active microwave sensors (radar): Advanced SCATterometer (ASCAT) aboard the EUMETSAT MetOp satellite

- Thermal sensors: Data from the thermal bands of the MODIS sensor onboard Terra satellite have also been used for retrieving soil moisture data.

Use of hyper-spectral remote sensing technique has been recently employed to improve the soil moisture simulation. Hyper-spectral monitoring of the soil moisture uses reflectivity in the VIS and the NIR bands to identify the changes in the spectral reflectance curves due to the presence of soil moisture (Yanmin et al., 2010). Spectral reflectance measured in multiple narrow bands in the hyperspectral image helps to extract most appropriate bands for the soil moisture estimation, and to identify the changes in the spectral reflectance curves due to the presence of soil moisture.

Watershed characterization and Prioritization

Watershed characterization involves the measurement and analysis of various hydro-geological and geo-morphological parameters, soil and land use characteristics etc. (Rao and Raju, 2010).

Watershed prioritization is the ranking of different watersheds or sub-watersheds within a watershed for any specific application based on the watershed characteristics.

Water Conservation and Rainwater Harvesting

Rainwater harvesting, wherein water from the rainfall is stored for future usage, is an effective water conservation measure particularly in the arid and semi-arid regions.

Rainwater harvesting techniques are highly location specific. Selection of appropriate water harvesting technique requires extensive field analysis to identify the rainwater harvesting potential of the area, and the physiographic and terrain characteristics of the locations. It depends on the amount of rainfall and its distribution, land topography, soil type and depth, and local socio-economic factors (Rao and Raju, 2010).

Rao and Raju (2010) had listed a set of parameters which need to be analyzed to fix appropriate locations for the water harvesting structures. These are:

- Rainfall

- Land use or vegetation cover

- Topography and terrain profile

- Soil type & soil depth

- Hydrology and water resources

- Socio-economic and infrastructure conditions

- Environmental and ecological impacts

Remote sensing techniques had been identified as potential tools to generate the basic information required for arriving at the most appropriate methods for each area.

In remote sensing aided analysis, various data layers were prepared and brought into a common GIS framework. Further, multi-criteria evaluation algorithms were used to aggregate the information from the basic data layers. Various decision rules were evaluated to arrive at the most appropriate solution as shown in the figure.

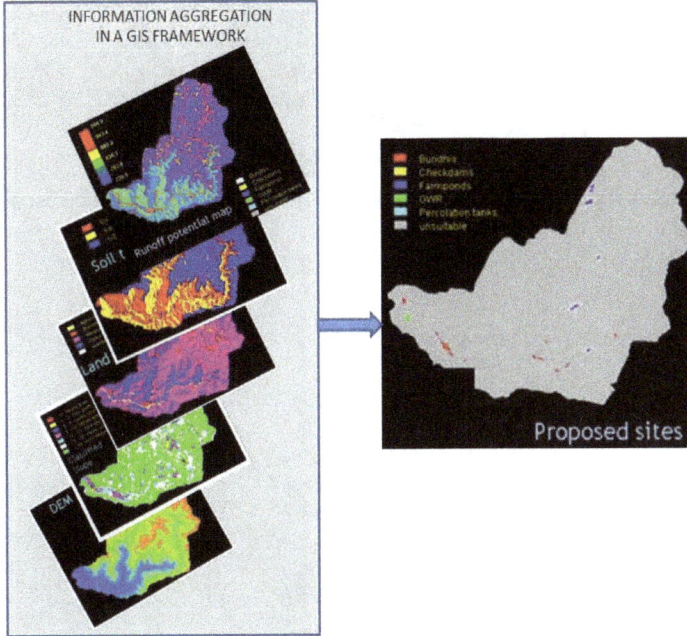

Schematic representation showing the remote sensing data aggregation in evaluating the suitability of various water harvesting techniques

The capability to provide large areal coverage at a fine spatial resolution makes remote sensing techniques highly advantageous over the conventional field-based surveys.

Runoff Model (Reservoir)

A runoff model is a mathematical model describing the rainfall–runoff relations of a rainfall *catchment area*, drainage basin or *watershed*. More precisely, it produces a surface runoff hydrograph in response to a rainfall event, represented by and input as a hyetograph. In other words, the model calculates the conversion of rainfall into runoff. A well known runoff model is the *linear reservoir*, but in practice it has limited applicability. The runoff model with a *non-linear reservoir* is more universally applicable, but still it holds only for catchments whose surface area is limited by the condition that the rainfall can be considered more or less uniformly distributed over the area. The maximum size of the watershed then depends on the rainfall characteristics of the region. When the study area is too large, it can be divided into sub-catchments and the various runoff hydrographs may be combined using flood routing techniques.

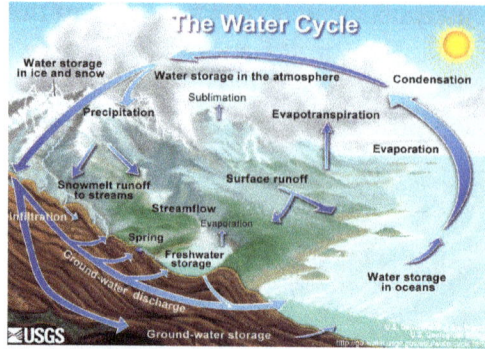

Rainfall-runoff models need to be calibrated before they can be used.

Linear Reservoir

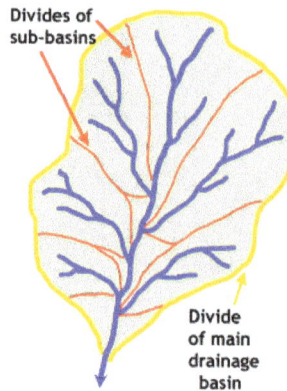

A watershed or drainage basin

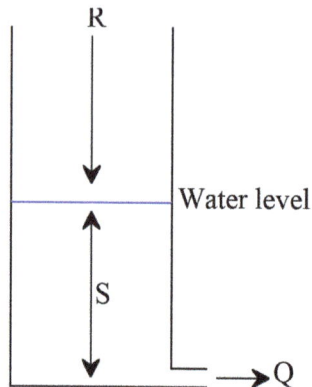

A linear reservoir

The hydrology of a linear reservoir is governed by two equations.

1. flow equation: $Q = A.S$, with units $[L/T]$, where L is length (e.g. mm) and T is time (e.g. h, day)

2. continuity or water balance equation: $R = Q + dS/dT$, with units $[L/T]$

where:

Q is the *runoff* or *discharge*

R is the *effective rainfall* or *rainfall excess* or *recharge*

A is the constant *reaction factor* or *response factor* with unit [1/T]

S is the water storage with unit [L]

dS is a differential or small increment of S

dT is a differential or small increment of T

Runoff Equation

A combination of the two previous equations results in a differential equation, whose solution is:

- $Q2 = Q1 \exp\{-A(T2 - T1)\} + R[1 - \exp\{-A(T2 - T1)\}]$

This is the *runoff equation* or *discharge equation*, where Q1 and Q2 are the values of Q at time T1 and T2 respectively while T2−T1 is a small time step during which the recharge can be assumed constant.

Computing the Total Hydrograph

Provided the value of A is known, the *total hydrograph* can be obtained using a successive number of time steps and computing, with the *runoff equation*, the runoff at the end of each time step from the runoff at the end of the previous time step.

Unit Hydrograph

The discharge may also be expressed as: $Q = -dS/dT$. Substituting herein the expression of Q in equation (1) gives the differential equation dS/dT = A.S, of which the solution is: $S = \exp(-A.t)$. Replacing herein S by Q/A according to equation (1), it is obtained that: $Q = A \exp(-A.t)$. This is called the instantaneous unit hydrograph (IUH) because the Q herein equals Q2 of the foregoing runoff equation using $R = 0$, and taking S as *unity* which makes Q1 equal to A according to equation (1).

The availability of the foregoing *runoff equation* eliminates the necessity of calculating the *total hydrograph* by the summation of partial hydrographs using the *IUH* as is done with the more complicated convolution method.

Determining the Response Factor A

When the *response factor* A can be determined from the characteristics of the watershed (catchment area), the reservoir can be used as a *deterministic model* or *analytical model*, see hydrological modelling. Otherwise, the factor A can be determined from a data record of rainfall and runoff using the method explained below under *non-linear reservoir*. With this method the reservoir can be used as a black box model.

Conversions

1 mm/day corresponds to 10 m³/day per ha of the watershed

1 l/s per ha corresponds to 8.64 mm/day or 86.4 m³/day per ha

Non-linear Reservoir

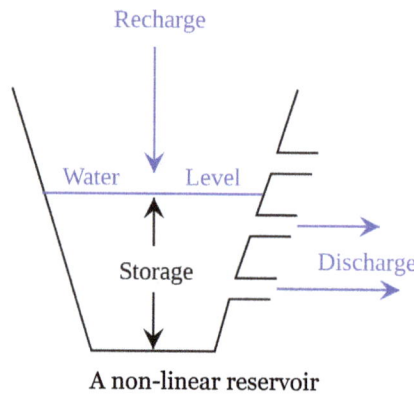

A non-linear reservoir

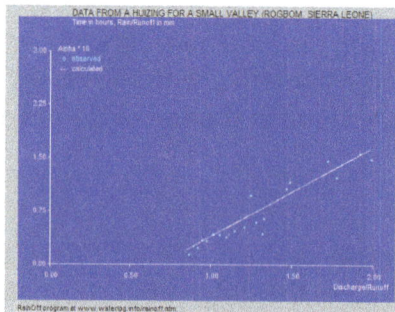

FThe reaction factor (Aq, Alpha) versus discharge (Q) for a small valley
(Rogbom) in Sierra Leone

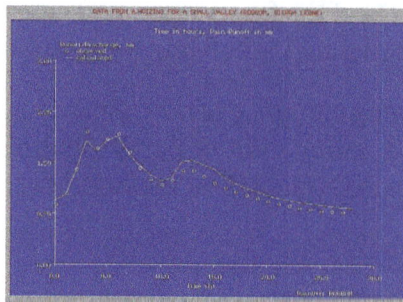

Actual and simulated discharge, Rogbom valley

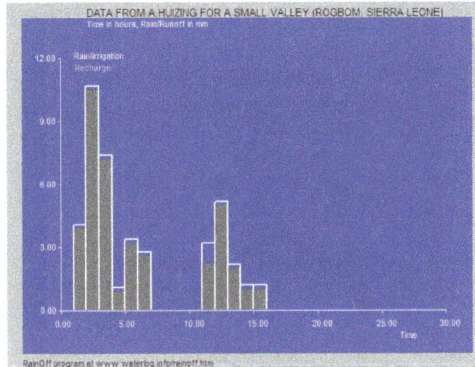

Rainfall and recharge, Rogbom valley

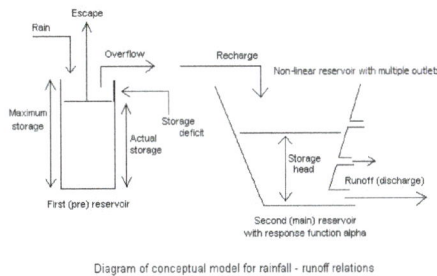

Diagram of conceptual model for rainfall - runoff relations

Non-linear reservoir with pre-reservoir for recharge

Contrary to the linear reservoir, the non linear reservoir has a reaction factor A that is not a constant, but it is a function of S or Q.

Normally A increases with Q and S because the higher the water level is the higher the discharge capacity becomes. The factor is therefore called Aq instead of A. The non-linear reservoir has *no* usable unit hydrograph.

During periods without rainfall or recharge, i.e. when $R = 0$, the runoff equation reduces to

- $Q_2 = Q_1 \exp \{ - Aq (T_2 - T_1) \}$, or:

or, using a *unit time step* ($T_2 - T_1 = 1$) and solving for Aq:

- $Aq = - \ln (Q_2/Q_1)$

Hence, the reaction or response factor Aq can be determined from runoff or discharge measurements using *unit time steps* during dry spells, employing a numerical method.

The figure shows the relation between Aq (Alpha) and Q for a small valley (Rogbom) in Sierra Leone.

The figure shows observed and *simulated* or *reconstructed* discharge hydrograph of the watercourse at the downstream end of the same valley.

Recharge

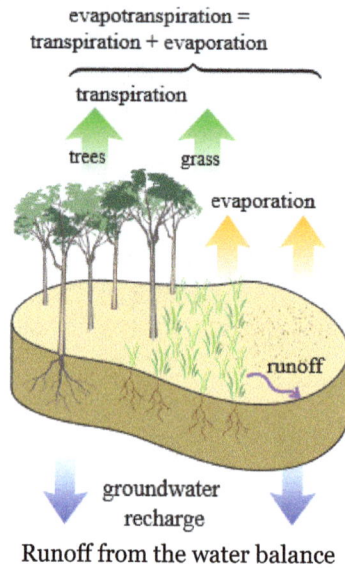

Runoff from the water balance

The recharge, also called *effective rainfall* or *rainfall excess*, can be modeled by a *pre-reservoir* giving the recharge as *overflow*. The pre-reservoir knows the following elements:

- a maximum storage (Sm) with unit length [L]

- an actual storage (Sa) with unit [L]

- a relative storage: Sr = Sa/Sm

- a maximum escape rate (Em) with units length/time [L/T]. It corresponds to the maximum rate of evaporation plus *percolation* and groundwater recharge, which will not take part in the runoff process.

- an actual escape rate: Ea = Sr.Em

- a storage deficiency: Sd = Sm + Ea − Sa

The recharge during a unit time step (T2−T1=1) can be found from R = Rain − Sd The actual storage at the end of a *unit time step* is found as Sa2 = Sa1 + Rain − R − Ea, where Sa1 is the actual storage at the start of the time step.

The Curve Number method (CN method) gives another way to calculate the recharge. The *initial abstraction* herein compares with Sm − Si, where Si is the initial value of Sa.

Software

Figures were made with the RainOff program, designed to analyse rainfall and runoff using the non-linear reservoir model with a pre-reservoir. The program also contains an example of the hydrograph of an agricultural subsurface drainage system for which the value of A can be obtained from the system's characteristics.

The SMART hydrological model includes agricultural subsurface drainage flow, in addition to soil and groundwater reservoirs, to simulate the flow path contributions to streamflow.

V*flo* is another software program for modeling runoff. V*flo* uses radar rainfall and GIS data to generate physics-based, distributed runoff simulation.

The WEAP (Water Evaluation And Planning) software platform models runoff and percolation from climate and land use data, using a choice of linear and non-linear reservoir models.

The RS MINERVE software platform simulates the formation of free surface run-off flow and its propagation in rivers or channels. The software is based on object-oriented programming and allows hydrologic and hydraulic modeling according to a semi-distributed conceptual scheme with different rainfall-runoff model such as HBV, GR4J, SAC-SMA or SOCONT.

SWAT Model

SWAT (Soil & Water Assessment Tool) is a river basin scale model developed to quantify the impact of land management practices in large, complex watersheds. SWAT is a public domain software enabled model actively supported by the USDA Agricultural Research Service at the Blackland Research & Extension Center in Temple, Texas, USA. It is a hydrology model with the following components: weather, surface runoff, return flow, percolation, evapotranspiration, transmission losses, pond and reservoir storage, crop growth and irrigation, groundwater flow, reach routing, nutrient and pesticide loading, and water transfer. SWAT can be considered a watershed hydrological transport model. This model is used worldwide and is continuously under development. As of July 2012, more than 1000 peer-reviewed articles have been published that document its various applications.

Model Operation

SWAT is a continuous time model that operates on a daily time step at basin scale. The objective of such a model is to predict the long-term impacts in large basins of management and also timing of agricultural practices within a year (i.e., crop rotations, planting and harvest dates, irrigation, fertilizer, and pesticide application rates and timing). It can be used to simulate at the basin scale water and nutrients cycle in landscapes whose dominant land use is agriculture. It can also help in assessing the environmental efficiency of best management practices and alternative management policies. SWAT uses a two-level dissagregation scheme; a preliminary subbasin identification is carried out based on topographic criteria, followed by further discretization using land use and soil type considerations. Areas with the same soil type and land use form a Hydrologic Response Unit (HRU), a basic computational unit assumed to be homogeneous in hydrologic response to land cover change.

Irrigation Management

Remote Sensing for Irrigation Management

In irrigation management, remote sensing is used as a tool to collect spatial and temporal variations in the hydro-meteorological parameters, crop characteristics and soil characteristics. Some of the important applications of remote sensing in irrigation management are listed below.

- Assessment of water availability in reservoirs for optimal management of water to meet the irrigation demand

- Identifying, inventorying and assessment of irrigated crops

- Determination of irrigation water demand over space and time

- Distinguishing land irrigated by surface water bodies and by ground water withdrawals

- Estimation of crop yield

- Study on water logging and salinity problems in irrigated lands

- Irrigation scheduling based on water availability and water demand

- Evapotranspiration studies

- Irrigation system performance evaluation

Crop Classification and Identification of the Irrigated Areas

Crop classification using the satellite remote sensing images is one of the most common applications of remote sensing in agriculture and irrigation management. Multiple images corresponding to various cropping periods are generally used for this purpose. The spectral reflectance values observed in various bands of the images are related to specific crops with the help of ground truth data. Also, satellite images of frequent time intervals are used to capture the temporal variations in the spectral signature, using which the crop stages are identified. The table gives a sample list of spectral signatures, observed in the standard FCC, for different crops during different growth stages.

Table: Spectral signatures of different crops in different growth stages

Crop Type	Growth Stage at the time of Satellite Data Acquisition	Possible signature on a Standard FCC
Paddy	2 to 3 weeks after transplantation	Greenish black to Reddish black
Paddy	Peak vegetative phase	Dark Red
Groundnut	Peak vegetative phase	Shades of bright red
Sugarcane	Peak vegetative phase	Light Pink to Pink
Cotton	Peak vegetative phase	Pink to Red

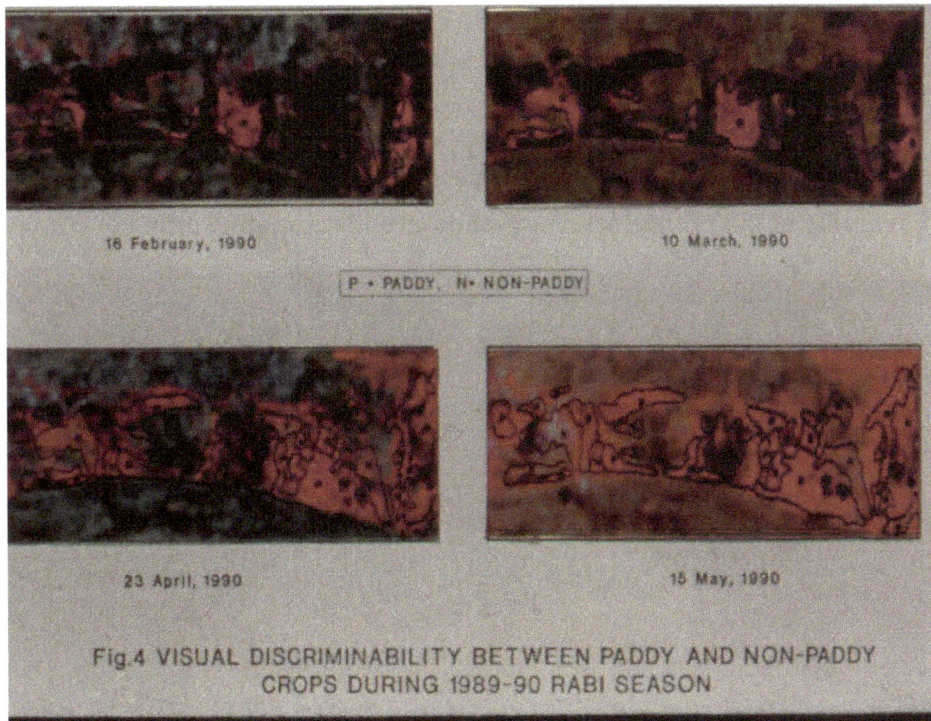

16 February, 1990   10 March, 1990

P • PADDY, N• NON-PADDY

23 April, 1990   15 May, 1990

Fig.4 VISUAL DISCRIMINABILITY BETWEEN PADDY AND NON-PADDY CROPS DURING 1989-90 RABI SEASON

Example for the crop type classification using remote sensing images

Identification of the irrigated area from the satellite images is based on the assessment of the crop health (using vegetation indices such as Normalized Differential Vegetation Index, NDVI) and the soil moisture condition. Irrigation water demand of the crops is defined by the actual evapotranspiration (ET) and the soil moisture availability. Remote sensing application in the estimation of irrigation water demand employs the estimation of ET by using the plant bio-physical parameters and the atmospheric parameters, and the soil moisture condition. The table gives a list of crop bio-physical parameters and their application in irrigation management. The following table lists the capability of the remote sensing techniques in estimating these parameters.

Table:List of crop bio-physical parameters and their application in irrigation management

Crop Parameter	Process	Purpose
Vegetation cover	Chorophyll development, soil and canopy fluxes	Irrigation area
Leaf area index	Biomass, minimum canopy resistance, heat fluxes	Yield, water use, water needs
Photosynthetically active radiation	Photosynthesis	Yield
Surface roughness	Aerodynamic resistance	Water use water needs
Surface albedo	Net radiation	Water use water needs
Thermal infrared surface emissivity	Net radiation	Water use water needs
Surface temperature	Net radiation, surface resistance	Water use
Surface resistance	Soil moisture and salinity	Water use
Crop coefficients	Grass evapotranspiration	Water needs
Transpiration coefficients	Potential soil and crop evaporation	Water use water needs
Crop yield	Accumulated biomass	Production

Table: Capability of the remote sensing techniques in estimating crop bio-physical parameters

Parameter	Accuracy	Need for field Data
Vegetation cover	High	None
Leaf area index	Good	None
Photosynthetically active radiation	Good	None
Surface roughness momentum	High	None
Surface roughness heat	Low	High
Surface albedo	Good	Low
Thermal infrared surface emissivity	Good	None
Surface temperature	Good	Low
Surface resistance	Good	None
Crop coefficients: tabulated	Moderate	None
Crop coefficients: analytical	Moderate	High
Transpiration coefficients	Good	None

Performance of the irrigation system is generally evaluated using indices such as relative water supply and relative irrigation supply. Bastiaanssen et al. (1998) has listed a set of irrigation performance indices derived with the help of the remote sensing data. Soil-adjusted vegetation index (SAVI), NDVI, transformed vegetation index (TVI), normalized difference wetness index (NDWI), green vegetation index (GVI) are a few of them.

Flood Mapping

Role of Remote Sensing Data in Flood Analyses

Remote sensing facilitates the flood surveys by providing the much needed information for flood studies. Satellite images acquired in different spectral bands during, before and after a flood event can provide valuable information about flood occurrence, intensity and progress of flood inundation, river course and its spill channels, spurs and embankments affected/ threatened etc. so that appropriate flood relief and mitigation measures can be planned and executed in time.

Poor weather condition generally associated with the floods, and poor accessibility due to the flooded water makes the ground and aerial assessment of flood inundated areas a difficult task. Application of satellite remote sensing helps to overcome these limitations. Through the selection of appropriate sensors and platforms, remote sensing can provide accurate and timely estimation of flood inundation, flood damage and flood-prone areas.

A list of sensors used for flood analyses are given in the table below.

Table: List of satellite sensors with their use for flood monitoring (Bhanumurthy et al., 2010)

Sl No:	Satellite	Sensor/ Mode	Spatial Resolution(m)	Spectral Resolution (μm)	Swath (km)	Used For
1.	IRS-P6	AWiFS	56	B2 : 0.52-0.59 B3 : 0.62-0.68 B4 : 0.77-0.86 B5 : 1.55-1.70	740	Regional level flood mapping
2.	IRS-P6	LISS-III	23.5	B2 : 0.52-0.59 B3 : 0.62-0.68 B4 : 0.77-0.86 B5 : 1.55-1.70	141	District-level flood mapping
3.	IRS-P6	LISS-IV	5.8 at nadir	B2 : 0.52-0.59 B3 : 0.62-0.68 B4:0.77-0.86	23.9	Detailed Mapping
4.	IRS-1D	WiFS	188	B3: 0.62-0.68 B4 : 0.77-0.86	810	Regional level flood mapping
5.	IRS-1D	LISS-III	23.5	B2: 0.52-0.59 B3 : 0.62-0.68 B4: 0.77-0.86 B5:1.55-1.70	141	Detailed Mapping
6.	Aqua / Terra	MODIS	250	36 in visible NIR & thermal	2330	Regional level Mapping

7.	IRS-P4	OCM	360	Eight narrow bands in visible & NIR	1420	Regional level Mapping
8.	Cartosat-1	PAN	2.5	0.5-0.85	30	Detailed Mapping
9.	Cartosat-2	PAN	1	0.45-0.85	9.6	Detailed Mapping
10.	Radarsat-1	SAR/ ScanSAR Wide	100	C-band (5.3 cm; HH Polarization)	500	Regional level mapping
11.	Radarsat-1	SAR /ScanSAR Narrow	50	C-band (5.3 cm)	300	District-level mapping
12	Radarsat-1	Standard	25	C-band	100	District-level mapping
13	Radarsat-1	Fine beam	8	C-band (5.3 cm)	50	Detailed mapping
14	Radarsat-2	SAR	3m ultra-find mode and 10m multi-llik fine mode	C –band	20 in ultra fine mode	Detailed mapping
14	ERS	SAR	25	C-band ; VV Polarization	100	District-level mapping

Remote Sensing Applications in Flood Analysis

The following 5 areas of remote sensing data application in flood analysis are identified:

- Flood mapping

- Near real-time monitoring of floods

- Flood damage assessment

- Flood hazard mapping

- River studies: mapping of river bank erosion and river course change

Environmental Monitoring

Remote Sensing in Water Quality Monitoring

The term water quality indicates the physical, chemical and biological characteristics of water. Temperature, chlorophyll content, turbidity, clarity, total suspended solids (TSS), nutrients, colored dissolved organic matter (CDOM), tripton, dissolved oxygen, pH, biological oxygen demand (BOD), chemical oxygen demand (COD), total organic carbon, and bacteria content are some of the commonly used water quality parameters.

In remote sensing, water quality parameters are estimated by measuring changes in the optical properties of water caused by the presence of the contaminants. Therefore, optical remote sensing has been commonly used for estimating the water quality parameters.

The following figure shows the Landsat TM image of the Fitzroy Estuary and Keppal Bay in Australia. The image taken on May 2003 shows the color difference of the water near the estuary mouth, which is due to the presence of suspended sediments.

Landsat TM image of the Fitzroy Estuary and Keppal Bay
in Australia in May 2003

Water quality parameters that have been successfully extracted using remote sensing techniques include chlorophyll content, turbidity, secchi depth, total suspended solids, colored dissolved organic matter and tripton. Thermal pollution in lakes and estuaries is monitored using thermal remote sensing techniques.

Algorithms for the Estimation of Water Quality Parameters from Remote Sensing Data

Estimation of water quality parameters using remote sensing data is based on the relationship between the concentration of the pollutant in the water and the consequent changes in the optical properties as observed in the satellite image.

Wavelengths or Bands used for Water Quality Monitoring

Optimum wavelength for monitoring water quality parameter through remote sensing depends on the substance that is measured.

Based on several *in-situ* analyses, the VIS and NIR portions of the EMR spectrum with wavelengths ranging from 0.7 to 0.8 μm have been considered to be the most useful bands for monitoring suspended sediments in water.

Optical remote sensing using the VIS and NIR bands has been preferred for measuring Chlorophyl content, turbidity, CDOM, Tripton etc.

Algorithms used for the Estimation of Water Quality Parameters

Algorithms or models used for the estimation of water quality parameters can be classified into Empirical relationships, Radiative transfer models or Physical models.

Empirical models use the relationship between the water quality parameter and the spectral records. General forms of such relationships are the following (Schmugge et al., 2002).

$$Y = A + BX \quad or \quad Y = AB^x \qquad (1)$$

where Y is the measurement obtained using the remote sensors and X is the water quality parameter of interest, and A and B are the empirical factors.

For example, an empirical relationship for estimating Chlorophyl content in water was given as follows (Harding et al., 1995)

$$Log_{10}[Chlorophyll] = A + B(-log_{10} G) \qquad (2)$$

$$G = \frac{(R_2)^2}{R_1.R_3} \qquad (3)$$

where A and B are empirical constants derived from *in situ* measurements, R_1, R_2 and R_3 are the radiances at 460 nm, 490 nm and 520 nm, respectively.

Similarly, Eq. 4 shows the empirical relationship for TSS. The algorithm is used to detect the TSS in water using the MODIS data. It is also known as TSM Clark algorithm.

$$TSM = 10^{\sum\limits_{i=0}^{5} a_i x^i}$$

Where $x = \log(nLw_1 + nLw_2 / nLw_4)$

and a_i = {0.490330, -2.712882, 3.412666, -8.336478, 12.111023, -5.961926}.

(4)

where nLw1 and nLw2 and nLw4 are the normalized water-leaving radiances on the dark blue band, second blue band and green band, respectively. These are related to the subsurface irradiance reflectance R (For more details refer Brando and Decker, 2003). It is to be noted that TSS stands for Total Suspended Solids and TSM is an acronym for Total Suspended Matter. The equation (4) represents one of the TSM models which are empirical in nature. Relations such as equation (4) can also be developed for TSS using bands of MODIS imageries.

Such relationships, based on field observations of the water quality parameters and the corresponding measurements obtained using the sensor, are controlled by the properties of water such as density, temperature etc. Therefore, the relationship derived for one field condition may not be valid for the other areas.

Radiative transfer models use a more general approach. Simplified solutions of the radiative transfer equations (RTEs) are used to relate the water surface reflectance (*Rrs*) to the controlling physical factors.

A sample RTE to relate the reflectance measured using remote sensing techniques to the suspended particulate matter is given below (Volpe et al., 2011)

$$R_{rs} = \frac{0.5 r_{rs}}{1 - 1.5 r_{rs}} \qquad (5)$$

$$r_{rs} = r_{rs}^{dp} [1 - e^{-(K_d + K_u^C)H}] + \frac{\rho_b}{\pi} e^{-(K_d + K_u^B)H} \qquad (6)$$

where

r_{rs} = subsurface remote sensing reflectance

r_{rs}^{dp} = r_{rs} for optically deep waters = $(0.084 + 0.17 u)u$

u = $b_b / (a + b_b)$, where b_b is the backscattering coefficient and a is the absorption coefficient

K_d = Vertically averaged diffuse attenuation coefficient for downwelling irradiance
= $Dd\ \alpha$

D_d = $1/cos(\theta w)$, where θw is the subsurface solar zenith angle

Ku^C = Vertically averaged diffuse attenuation coefficient for upwelling radiance from water column scattering = $Du^C \alpha$

Ku^B = Vertically average d diffuse attenuation coefficient for upwelling radiance from bottom reflectance = $Du^B \alpha$

α = $a + b_b$

Du^C = 1.03 $(1+2.4u)^{0.5}$

Du^B = 1.03 $(1+5.4u)^{0.5}$

ρb = Bottom albedo

H = water depth

The backscattering and the absorption coefficients were determined by calibration.

Satellites and Sensors used for Water Quality Monitoring

Remote sensing of the water quality parameter in the earlier days employed fine resolution optical images from the satellites e.g., Landsat TM. However, poor temporal coverage of the images (once in 16 days) was a major limitation in such studies. With the development of new satellites and sensors, the spatial, temporal and radiometric resolutions have improved significantly. Using sensors such as MODIS (with 36 spectral bands) and MERIS (with 15 spectral bands) better accuracy in the estimation of water quality parameters is now possible.

A recent development in the remote sensing application in water quality monitoring is the use of hyper-spectral images in monitoring the water quality parameters. The large number of narrow spectral bands used in the hyper-spectral sensors help in improved detection of the contaminants and the organic matters present in water. Use of hyper-spectral images to monitor the tropic status of lakes and estuaries, assessment of total suspended matter and chlorophyll content in the surface water and bathymetric surveys are a few examples.

For more details on the hyperspectral remote sensing data application in water quality monitoring, refer Koponen et al., 2002; Thiemann and Kaufmann, 2002; Hakvoort et al., 2002; Lesser and Mobley, 2007.

The following table gives a brief summary of some of the works wherein the remote sensing data have been used for estimating the water quality parameters.

The figures show the application of remote sensing data for monitoring various water quality parameters.

Table: Important water quality parameters estimated and the characteristics of the sensors used (Source: Nagesh Kumar and Reshmidevi, 2013)

Parameter	Sensor type	Sensor / data	Remote sensing data characteristics	Algorithm used	Reference
Chloro-phyll	MSS	MERIS	15 spectral bands, 300 m spatial resolution, poor temporal coverage	Spectral curves were calibrated using field observations	Koponen et al., 2002
				ESA BEAM tool box	Giardino et al., 2010
		Landsat TM	7 spectral bands, 30 m spatial resolution, poor temporal coverage	Empirical relation	Brezonik et al., 2005
		Sea-WiFS, MODIS	More number of spectral bands, 250- 1000 m spatial resolution, better temporal coverage,	Band ratio algorithm	Lesht et al., 2013
	Hyperspectral	Hyper-ion	Better spectral resolution, 30 m spatial resolution, poor temporal coverage	Analytical method, Numerical radiative transfer model	Brando et al., 2003
				Bio-optical model	Santini et al., 2010
CODM, Tripton	Hyperspectral	Hyper-ion	Better spectral resolution, 30 m spatial resolution, poor temporal coverage	Analytical method, Numerical radiative transfer model	Brando et al., 2003
				Bio-optical model	Santini et al., 2010
Secchi depth, Turbidity	MSS	MERIS	15 spectral bands, 300 m spatial resolution, poor temporal coverage	Spectral curves were calibrated using field observations	Schmugge et al., 1992
				ESA BASE toolbox	Koponen et al., 2002
		Landsat TM	7 spectral bands, 30 m spatial resolution, poor temporal coverage	Empirical relation	Brezonik et al., 2005
TSS	MSS	MERIS	15 spectral bands, 300 m spatial resolution, poor temporal coverage	ESA BASE tool box	Giardino et al., 2010
		Landsat TM	7 spectral bands, 30 m spatial resolution, poor temporal coverage	Empirical relation	Brezonik et al., 2005

| Surface temperature | Thermal | MODIS–LST | Better temporal coverage, 250-1000 m spatial resolution | MODIS Level-2 temperature data | Alcântara et al., 2010; Giardino et al., 2010 |
| | | AVHRR | 5 bands (3 thermal bands), good temporal coverage, 1000-2000 m spatial resolution | Multi-Channel SST estimation algorithm (MCSST) | Politi et al., 2012 |

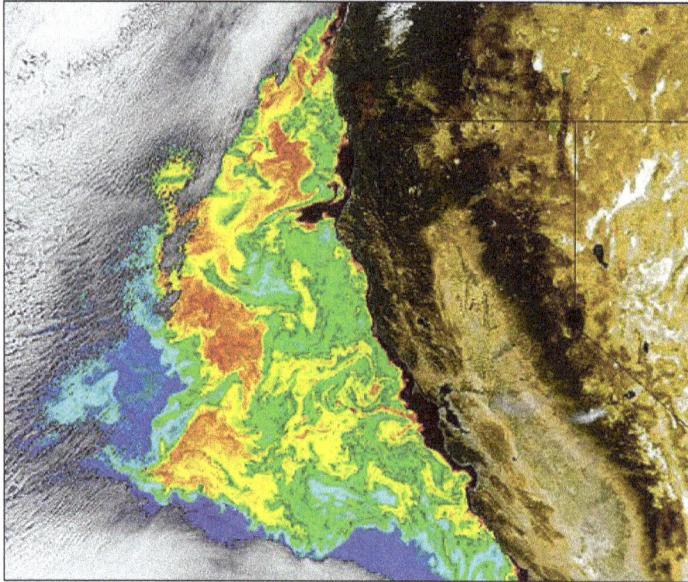

Chlorophyll concentration in the off-coast of California estimated using the SeaWiFS and MODIS sensors. Bright red indicates high concentration and blues indicate low concentrations

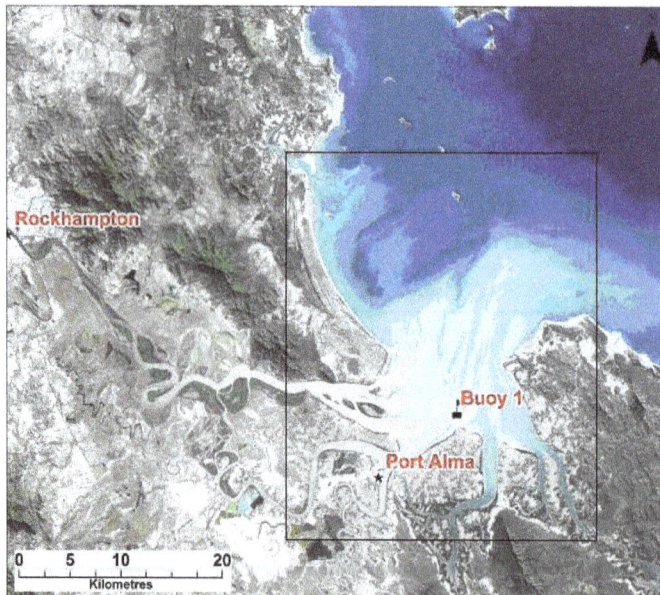

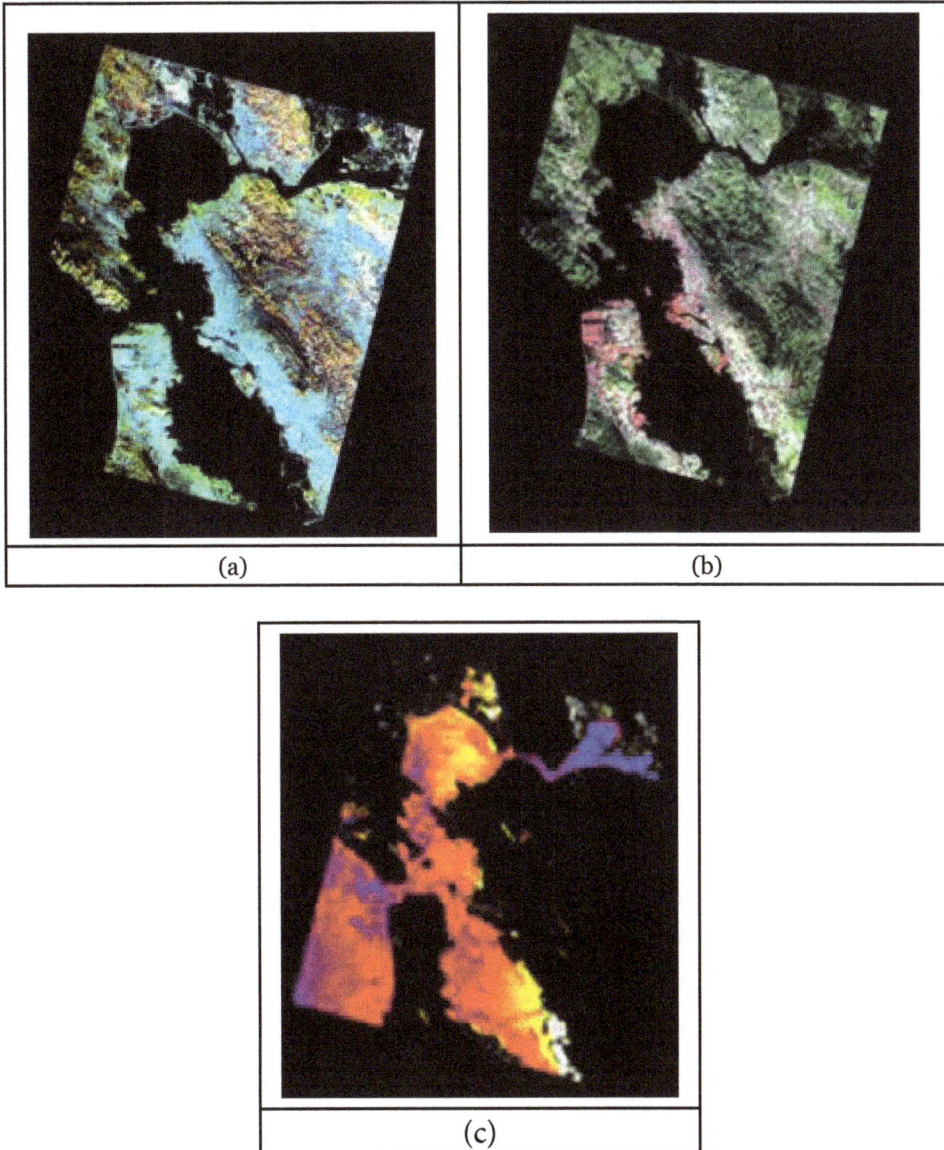

ASTER images of the San Francisco Bay area (a) From SWIR bands (b) A composite using thermal data and visible bands (c) Thermal data showing temperature variations only in water. Land areas are masked out.

In Fig.(c), colour varies from red for the warmest to blue for the coolest areas. The warmest temperatures are found in San Francisco and across the Bay in the Oakland group of cities, which may be mostly due to the thermal pollution from the large number of industries located in the area.

Remote Sensing Application in Monitoring Land Degradation

Land degradation is the deterioration of the land or soil properties that negatively affect the effective functioning of the land based ecosystems. From the agricultural perspective it may be defined as the reduction in soil capacity to produce crops. From the ecological perspective, land degradation causes damage to the healthy functioning of the land based ecosystems.

Land degradation may be either due to natural factors such as floods, drought, earthquake, or due to the human induced factors like over exploitation of land and water resources, or unscientific land use. The following are considered to be some of the major factors causing land degradation (Ravishankar and Sreenivas, 2010).

- Water erosion: Displacement of soil material by water

- Wind erosion: Displacement of top soil by wind

- Water logging: Extensive ponding for a long time affecting the productivity of the land

- Salinization: Chemical imbalance in the soil causing desiccation of the plants or non- availability of essential nutrients to plants

- Acidification: Increase in the hydrogen cations in the soil affecting the plant health

- Anthropogenic: Mining, industries leading to decreased productivity of the land

- Others: Barren areas, rocky waste areas, riverine sand areas, sea ingression areas etc.

Vast areas in the world are currently affected by land degradation. According to the Department of Land Resources, in 2005 around 55.27 million hectares of land in India is affected due to some sort of degradation . Scientific information about the degraded land, or rate of land degradation is necessary for land reclamation and management.

PERMISSIONS

All chapters in this book are published with permission under the Creative Commons Attribution Share Alike License or equivalent. Every chapter published in this book has been scrutinized by our experts. Their significance has been extensively debated. The topics covered herein carry significant information for a comprehensive understanding. They may even be implemented as practical applications or may be referred to as a beginning point for further studies.

We would like to thank the editorial team for lending their expertise to make the book truly unique. They have played a crucial role in the development of this book. Without their invaluable contributions this book wouldn't have been possible. They have made vital efforts to compile up to date information on the varied aspects of this subject to make this book a valuable addition to the collection of many professionals and students.

This book was conceptualized with the vision of imparting up-to-date and integrated information in this field. To ensure the same, a matchless editorial board was set up. Every individual on the board went through rigorous rounds of assessment to prove their worth. After which they invested a large part of their time researching and compiling the most relevant data for our readers.

The editorial board has been involved in producing this book since its inception. They have spent rigorous hours researching and exploring the diverse topics which have resulted in the successful publishing of this book. They have passed on their knowledge of decades through this book. To expedite this challenging task, the publisher supported the team at every step. A small team of assistant editors was also appointed to further simplify the editing procedure and attain best results for the readers.

Apart from the editorial board, the designing team has also invested a significant amount of their time in understanding the subject and creating the most relevant covers. They scrutinized every image to scout for the most suitable representation of the subject and create an appropriate cover for the book.

The publishing team has been an ardent support to the editorial, designing and production team. Their endless efforts to recruit the best for this project, has resulted in the accomplishment of this book. They are a veteran in the field of academics and their pool of knowledge is as vast as their experience in printing. Their expertise and guidance has proved useful at every step. Their uncompromising quality standards have made this book an exceptional effort. Their encouragement from time to time has been an inspiration for everyone.

The publisher and the editorial board hope that this book will prove to be a valuable piece of knowledge for students, practitioners and scholars across the globe.

Index

www.ingramcontent.com/pod-product-compliance
Lightning Source LLC
Chambersburg PA
CBHW061952190326
41458CB00009B/2856